Robert Cochrane

Heroes of Invention and Discovery

Lives of Eminent inventors and pioneers in science

Robert Cochrane

Heroes of Invention and Discovery
Lives of Eminent inventors and pioneers in science

ISBN/EAN: 9783337193157

Printed in Europe, USA, Canada, Australia, Japan

Cover: Foto ©Andreas Hilbeck / pixelio.de

More available books at **www.hansebooks.com**

RISEN BY PERSEVERANCE;

OR,

LIVES OF SELF-MADE MEN.

Compiled and Arranged

BY

ROBERT COCHRANE,

EDITOR OF

'THE ENGLISH ESSAYISTS,' 'TREASURY OF BRITISH ELOQUENCE,' 'TREASURY OF MODERN BIOGRAPHY,' 'THE ENGLISH EXPLORERS,' ETC. ETC.

EDINBURGH:
W. P. NIMMO, HAY, & MITCHELL.

PREFATORY NOTE.

THE series of popular biographies in which the present volume is included, is issued by the publishers with the view of meeting the steadily increasing demand for biographical reading of a wholesome and instructive character. To those who have afterwards to mingle actively in practical life, there is always something stimulating in noting how those who have gone before—men eminent in many departments—may have conducted themselves in the battle of life. Perhaps as in everyday life, example is here also more powerful than precept; the less moralizing the better, the influence and practical lesson from each career coming home almost insensibly, yet none the less powerfully, to the reader. The two first sketches in the book are quoted from a well-known work, the late Professor Craik's *Pursuit of Knowledge under Difficulties*, London, 1830; *William Cobbett* appeared many years ago in *Tait's Magazine*. The remaining articles have been specially prepared or selected for the present volume.

CONTENTS.

	PAGE
BENJAMIN FRANKLIN,	5
JAMES BRINDLEY,	51
WILLIAM COBBETT,	73
HUGH MILLER,	104
SIR TITUS SALT,	140
CHARLES DICKENS,	172

RISEN BY PERSEVERANCE.

BENJAMIN FRANKLIN.

THE name we are now to mention is perhaps the most distinguished to be found in the annals of self-education. Of all those, at least, who, by their own efforts, and without any usurpation of the rights of others, have raised themselves to a high place in society, there is no one, as has been remarked, the close of whose history presents so great a contrast to its commencement as that of BENJAMIN FRANKLIN. It fortunately happens, too, in his case, that we are in possession of abundant information as to the methods by which he contrived to surmount the many disadvantages of his original condition; to raise himself from the lowest poverty and obscurity to affluence and distinction; and, above all, in the absence of instructors, and of the ordinary helps to the acquisition of knowledge, to enrich himself so plentifully with the treasures of literature and science, as not only to be enabled to derive from that source the chief happiness of his life, but to succeed in placing himself high among the most famous writers and philosophers of his time.

Franklin has himself told us the story of his early life inimitably well. The narrative is given in the form of a letter to his son, and does not appear to have been written originally with any view to publication. 'From the poverty and obscurity,' he says, 'in which I was born, and in which I passed my earliest years, I have raised myself to a state of affluence, and some degree of celebrity in the world. As constant good fortune has accompanied me, even to an advanced period of life, my posterity will perhaps be desirous of learning the means which I employed, and which, thanks to Providence, so well succeeded with me. They may also deem them fit to be imitated, should any of them find themselves in similar circumstances.' It is now many years (1817) since this letter was, for the first time, given to the world by the grandson of the illustrious writer, William Temple Franklin, only a small portion of it having previously appeared, and that merely a re-translation into English from a French version of the original manuscript which had been published at Paris, and which is not wholly trustworthy.

Franklin was born at Boston, in North America, on the 17th of January 1706; the youngest, with the exception of two daughters, of a family of seventeen children. His father, who had emigrated from England about twenty-four years before, followed the occupation of a soap-boiler and tallow-chandler,—a business to which he had not been bred, and by which he seems with difficulty to have been able to support his numerous family. At first it was proposed to make Benjamin a clergyman; and he was accordingly, having before learned to read, put to the grammar-school at eight years of age,—an uncle, whose namesake he was, and who appears to have been an ingenious man, encouraging the project, by offering to give him several volumes of sermons to set up with, which he had taken down, in a shorthand of his own invention, from the different preachers he had been in the habit of hearing. This person, who was now advanced in life, had been only a common silk-dyer, but had been both a great reader and writer in his day, having filled two quarto volumes with his own manuscript poetry. What he was most proud of, however, was his shorthand, which he was very anxious that his nephew should learn. But young Franklin had not been quite a year at the grammar-school, when his father began to reflect that the expense of a college education for him was what he could not very well afford; and that, besides, the church in America was a poor profession after all. He was accordingly removed, and placed for another year under a teacher of writing and arithmetic; after which his father took him home, when he was no more than ten years old, to assist him in his own business. Accordingly, he was employed, he tells us, in cutting wicks for the candles, filling the moulds for cast candles, attending the shop, going errands, and other drudgery of the same kind. He showed so much dislike, however, to this business, that his father, afraid he would break loose and go to sea, as one of his elder brothers had done, found it advisable, after a trial of two years, to look about for another occupation for him; and taking him round to see a great many different sorts of tradesmen at their work, it was at last agreed upon that he should be bound apprentice to a cousin of his own, who was a cutler. But he had been only for some days on trial at this business, when, his father thinking the apprentice fee which his cousin asked too high, he was again taken home. In this state of things it was finally resolved to place him with his brother James, who had been bred a printer, and had just returned from England and set up on his own account at Boston. To him, therefore, Benjamin was bound apprentice, when he was yet only in his twelfth year, on an agreement that he should remain with him in that capacity till he reached the age of twenty-one.

One of the principal reasons which induced his father to determine upon this profession for him was the fondness he had from his infancy shown for reading. All the money he could get hold of used to be eagerly laid out in the purchase of books. His father's small collection consisted principally of works in controversial divinity, a subject of little interest to a reader of his age; but, such as they were, he went through most of them. Fortunately there was also a copy of *Plutarch's Lives*, which he says he read abundantly. This, and a book by Daniel Defoe, called *An Essay on Projects*, he seems to think were the two works from which he derived the most advantage. His new profession of a printer, by procuring him the acquaintance of some booksellers' apprentices, enabled him considerably to extend his acquaintance with books, by frequently borrowing a volume in the evening, which he sat up reading the greater part of the night, in order that he might return it in the morning, lest it should be missed. But these solitary studies did not prevent him from soon acquiring a great proficiency in his business, in which he was every day becoming more useful to his brother. After some time, too, his access to books was greatly facilitated by the kindness of a liberal-minded merchant who was in the habit of frequenting the printing office, and, being possessed of a tolerable library, invited young Franklin, whose industry and intelligence had attracted his attention, to come to see it; after which he allowed him to borrow from it such volumes as he wished to read.

Our young student was now to distinguish himself in a new character. The perusal of the works of others suggested to him the idea of trying his own talent at composition; and his first attempts in this way were a few pieces of poetry. Verse, it may be observed, is generally the earliest sort of composition attempted either by nations or individuals, and for the same reasons in both cases,—namely, first, because poetry has peculiar charms for the unripe understanding; and secondly, because people at first find it difficult to conceive what composition is at all, independently of such measured cadences and other regularities as constitute verse. Franklin's poetical fit, however, did not last long. Having been induced by his brother to write two ballads, he was sent to sell them through the streets; and one of them at least, being on a subject which had just made a good deal of noise in the place, sold, as he tells us, prodigiously. But his father, who, without much literary knowledge, was a man of a remarkably sound and vigorous understanding, soon brought down the rising vanity of the young poet, by pointing out to him the many faults of his performances, and convincing him what wretched stuff they really were. Having been told, too, that verse-makers were generally beggars, with his characteristic prudence he determined to write no more ballads.

He had an intimate acquaintance of the name of Collins, who was, like himself, passionately fond of books, and with whom he was in the habit of arguing upon such subjects as they met with in the course of their reading. Among other questions which they discussed in this way, one accidentally arose on the abilities of women, and the propriety of giving them a learned education. Collins maintained their natural unfitness for any of the severer studies, while Franklin took the contrary side of the question,—'perhaps,' he says, 'a little for dispute sake.' His antagonist had always the greater plenty of words; but Franklin thought that, on this occasion in particular, his own arguments were rather the stronger; and on their parting without settling the point, he sat down and put a summary of what he advanced in writing, which he copied out and sent to Collins. This gave a new form to the discussion, which was now carried on for some time by letters, of which three or four had been written on both sides, when the correspondence fell into the hands of Franklin's father. His natural acuteness and good sense enabled him here again to render an essential service to his son, by pointing out to him how far he fell short of his antagonist in elegance of expression, in method, and in perspicuity, though he had the advantage of him in correct spelling and punctuation, which he evidently owed to his experience in the printing-office. From that moment Franklin determined to spare no pains in endeavouring to improve his style; and we shall give, in his own words, the method he pursued for that end.

'About this time,' says he, 'I met with an odd volume of the *Spectator:* I had never before seen any of them. I bought it, read it over and over, and was much delighted with it. I thought the writing excellent, and wished, if possible, to imitate it. With that view I took some of the papers, and making short hints of the sentiments in each sentence, laid them by a few days, and then, without looking at the book, tried to complete the papers again by expressing each hinted sentiment at length, and as fully as it had been expressed before, in any suitable words that should occur to me. Then I compared my *Spectator* with the original, discovered some of my faults, and corrected them. But I found I wanted a stock of words, or a readiness in recollecting and using them, which I thought I should have acquired before that time if I had gone on making verses; since the continual search for words of the same import, but of different length to suit the measure, or of different sound for the rhyme, would have laid me under a constant necessity of searching for variety, and also have tended to fix that variety in my mind, and make me master of it. Therefore I took some of the tales in the *Spectator* and turned them into verse; and after a time, when I had pretty well forgotten the prose, turned them back again. I also sometimes jumbled my collection of hints into confusion, and, after some weeks, endeavoured to reduce them into the best order before I began to form the full sentences and complete the subject. This was to teach me method in the arrangement of the thoughts. By comparing my work with the original I discovered many faults and corrected them; but I sometimes had the pleasure to fancy that in certain particulars of small consequence I had been fortunate enough to improve the method or the language; and this encouraged me to think that I might in time come to be a tolerable English writer, of which I was extremely ambitious.'

Even at this early age nothing could exceed the perseverance and self-denial which he displayed in pursuing his favourite object of cultivating his mental faculties to the utmost of his power. When only sixteen, he chanced to meet with a book in recommendation of a vegetable diet, one of the arguments at least in favour of which made an immediate impression upon him,—namely, its greater cheapness; and from this and other considerations, he determined to adopt that way of living for the future. Having taken this resolution, he proposed to his brother, if he would give him weekly only half what his board had hitherto cost, to board himself, an offer which was immediately accepted. He presently found that by adhering to his new system of diet he could still save half what his brother allowed him. 'This,' says he, 'was an additional fund for buying of books; but I had another advantage in it. My brother and the rest going from the printing-house to their meals, I remained there alone, and despatching presently my light repast (which was often no more than a biscuit or a slice of bread, a handful of raisins or a tart from the pastrycook's, and a glass of water), had the rest of the time till their return for study; in which I made the greater progress, from that greater clearness of head and quicker apprehension which generally attend temperance in eating and drinking.' It was about this time that, by means of Cocker's Arithmetic, he made himself master of that science, which he had twice attempted in vain to learn while at school; and that he also obtained some acquaintance with the elements of geometry, by the perusal of a treatise on Navigation. He mentions, likewise, among the works which he now read, *Locke on the Human Understanding*, and the Port-Royal *Art of Thinking;* together with two little sketches on the arts of Logic and Rhetoric, which he found at the end of an English grammar, and which initiated him in the Socratic mode of disputation, or that way of arguing by which an antagonist, by being questioned, is imperceptibly drawn into admissions which are afterwards dexterously turned against him. Of this method of reasoning he became, he tells us, excessively fond, finding it very safe for himself, and very embarassing for those against whom he used it; but he afterwards abandoned it, apparently from a feeling that it gave advantages rather to cunning than to truth, and was better adapted to gain victories in conversation, than either to convince or to inform.

A few years before this, his brother had begun to publish a newspaper, the second that had appeared in America. This brought most of the literary people of Boston occasionally to the printing-office; and young Franklin often heard them conversing about the articles that appeared in the newspaper, and the approbation which particular ones received. At last, inflamed with the ambition of sharing in this sort of fame, he resolved to try how a communication of his own would succeed. Having written his paper, therefore, in a disguised hand, he put it at night under the door of the printing-office, where it was found in the morning, and submitted to the consideration of the critics, when they met as usual. 'They read it,' says he; 'commented on it in my hearing; and I had the exquisite pleasure of finding it met with their approbation; and that in their different guesses at the author, none were named but men of some character among us for learning and ingenuity.' 'I suppose,' he adds, 'that I was rather lucky in my judges, and that they were not really so very good as I then believed them to be.' Encouraged, however, by the success of this attempt, he sent several other pieces to the press in the same way, keeping his secret, till, as he expresses it, all his fund of sense for such performances was exhausted. He then discovered himself, and immediately found that he began to be looked upon as a person of some consequence by his brother's literary acquaintances.

This newspaper soon after afforded him, very unexpectedly, an opportunity of extricating himself from his indenture to his brother, who had all along treated him with great harshness, and to whom his rising literary reputation only made him more an object of envy and dislike. An article which they had admitted having offended the local government, his brother, as proprietor of the paper, was not only sentenced to a month's imprisonment, but prohibited from any longer continuing to print the offensive journal. In these circumstances, it was determined that it should appear for the future in the name of Benjamin, who had managed it during his brother's confinement; and in order to prevent it being alleged that the former proprietor was only screening himself behind one of his apprentices, the indenture by which the latter was bound was given up to him; he at the same time, in order to secure to his brother the benefit of his services, signing new indentures for the remainder of his time, which were to be kept private. 'A very flimsy scheme it was,' says Franklin; 'however, it was immediately executed; and the paper was printed accordingly under my name for several months. At length a fresh difference arising between my brother and me, I took upon me to assert my freedom, presuming that he would not venture to produce the new indentures. It was not fair in me to take this advantage, and this I therefore reckon one of the first *errata* of my life; but the unfairness of it weighed little with me, when under the impressions of resentment for the blows his passion too often urged him to bestow upon me, though he was otherwise not an ill-natured man: perhaps I was too saucy and provoking.'

Finding, however, that his brother, in consequence of this

exploit, had taken care to give him such a character to all those of his own profession in Boston, that nobody would emqloy him there, he now resolved to make his way to New York, the nearest place where there was a printer; and accordingly, after selling his books to raise a little money, he embarked on board a vessel for that city, without communicating his intention to his friends, who he knew would oppose it. In three days he found himself at the end of his voyage, near three hundred miles from his home, at the age of seventeen, without the least recommendation, as he tells us, or knowledge of any person in the place, and with very little money in his pocket. Worst of all, upon applying to the only printer likely to give him any employment, he found that this person had nothing for him to do, and that the only way in which he could serve him was by recommending him to proceed to Philadelphia, a hundred miles farther, where he had a son, who, he believed, might employ him. We cannot follow our runaway through the disastrous incidents of this second journey; but, for the reason which he states himself, we shall allow him to give his own most graphic description of his first appearance in Philadelphia.

After concluding the account of his voyage, 'I have been the more particular,' says he, 'in this description of my journey, and shall be so of my first entry into that city, that you may, in your mind, compare such unlikely beginnings with the figure I have since made there. I was in my working dress, my best clothes coming round by sea. I was dirty, from my being so long in the boat; my pockets were stuffed out with shirts and stockings; and I knew no one, nor where to look for lodging. Fatigued with walking, rowing, and the want of sleep, I was very hungry; and my whole stock of cash consisted in a single dollar, and about a shilling in copper coin, which I gave to the boatmen for my passage. At first they refused it, on account of my having rowed; but I insisted on their taking it. Man is sometimes more generous when he has little money than when he has plenty; perhaps to prevent his being thought to have but little. I walked towards the top of the street, gazing about till near Market Street, where I met a boy with bread. I had often made a meal of dry bread, and inquiring where he had bought it, I went immediately to the baker's he directed me to. I asked for biscuits, meaning such as we had at Boston; that sort, it seems, was not made in Philadelphia. I then asked for a threepenny loaf, and was told they had none. Not knowing the different prices, nor the names of the different sorts of bread, I told him to give me three penny-worth of any sort. He gave me, accordingly, three great puffy rolls. I was surprised at the quantity, but took it; and having no room in my pockets, walked off with a roll under each arm, and eating the other. Thus I went up Market Street, as far as Fourth Street, passing by the door of Mr. Read, my future wife's father, when she, standing at the door, saw me, and thought I made, as I certainly did, a most awkward, ridiculous appearance. Then I turned and went down Chesnut Street and part of Walnut Street, eating my roll all the way, and coming round, found myself again at Market Street Wharf, near the boat I came in, to which I went for a draught of the river water; and being filled with one of my rolls, gave the other two to a woman and her child that came down the river in the boat with us, and were waiting to go farther. Thus refreshed, I walked again up the street, which by this time had many clean dressed people in it, who were all walking the same way. I joined them, and thereby was led into the great meeting-house of the Quakers, near the market. I sat down among them; and after looking round a while, and hearing nothing said, being very drowsy, through labour and want of rest the preceding night, I fell fast asleep, and continued so till the meeting broke up, when some one was kind enough to rouse me. This, therefore, was the first house I was in, or slept in, in Philadelphia.'

Refreshed by his brief sojourn in this cheap place of repose, he then set out in quest of a lodging for the night. Next morning he found the person to whom he had been directed, who was not, however, able to give him any employment; but upon applying to another printer in the place, of the name of Keimer, he was a little more fortunate, being set by him, in the first instance, to put an old press to rights, and afterwards taken into regular work. He had been some months at Philadelphia, his relations in Boston knowing nothing of what had become of him, when a brother-in-law, who was the master of a trading sloop, happening to hear of him in one of his voyages, wrote to him in very earnest terms to entreat him to return home. The letter which he sent in reply to this application reaching his brother-in-law when he chanced to be in company with Sir William Keith, the Governor of the Province, it was shown to that gentleman, who expressed considerable surprise on being told the age of the writer; and immediately said that he appeared to be a young man of promising parts, and that if he would set up on his own account in Philadelphia, where the printers were wretched ones, he had no doubt he would succeed: for his part, he would procure him the public business, and do him every service in his power. Some time after this, Franklin, who knew nothing of what had taken place, was one day at work along with his master near the window, when 'we saw,' says he, 'the Governor and another gentleman (who proved to be Colonel French, of Newcastle, in the province of Delaware), finely dressed, come directly across the street to our house, and heard them at the door. Keimer ran down immediately, thinking it a visit to him; but the Governor inquired for me, came up, and with a condescension and politeness I had been quite unused to, made me many compliments, desired to be acquainted with me, blamed me kindly for not having made myself known to him when I first came to the place, and would have me away with him to the tavern, where he was going with Colonel French, to taste, as he said, some excellent Madeira. I was not a little surprised, and Keimer stared with astonishment.'

The reader already perceives that Sir William must have been rather an odd sort of person; and this becomes still more apparent in the sequel of the story. Having got his young protege to the tavern, he proposed to him, over their wine, that he should as soon as possible set up in Philadelphia as a master printer, only continuing to work with Keimer till an opportunity should offer of a passage to Boston, when he would return home, to arrange the matter with his father, who, the Governor had no doubt, would, upon a letter from him, at once advance his son the necessary funds for commencing business. Accordingly, Franklin set out for Boston by the first vessel that sailed; and, upon his arrival, was very kindly received by all his family, except his brother, and surprised his father not a little by presenting him with the Governor's letter. For some time his father said little or nothing on the subject, merely remarking, that Sir William must be a person of small discretion, to think of setting a youth up in business who wanted three years to arrive at man's estate. But at last he decidedly refused to have anything to do with the arrangement; and Franklin returned to his patron to tell him of his bad success, going this time, however, with the consent and blessing of his parents, who, finding how industrious he had been while in Philadelphia, were willing that he should continue there. When Franklin presented himself to Sir William with his father's answer to the letter he had been honoured with from that functionary, the Governor observed that he was too prudent: 'But since he will not set you up,' added he, 'I will do it myself.' It was finally agreed that Franklin should proceed in person to England, to purchase types and other necessary articles, for which the Governor was to give him letters of credit to the extent of one hundred pounds.

After repeated applications to the Governor for the promised letters of credit, Franklin was at last sent on board the vessel for England, which was just on the point of sailing, with an assurance that Colonel French should be sent to him with the letters immediately. That gentleman soon after made his appearance, bearing a packet of despatches from the Governor: in this packet Franklin was informed his letters were. Accordingly, when they got into the British Channel, the captain having allowed him to search for them among the others, he found several addressed to his care, which he concluded of course to be those he had been promised. Upon presenting one of them, however, to a stationer, to whom it was directed, the man, having opened it, merely said, 'Oh, this is from Riddlesdon (an attorney in Philadelphia, whom Franklin knew to be a thorough knave); I have lately found him to be a complete rascal;' and giving back the letter, turned on his heel, and proceeded to serve his customers. Upon this, Franklin's confidence in his patron began to be a little shaken; and, after reviewing the whole affair in his own mind, he resolved to lay it before a very intelligent mercantile gentleman, who had come over from America with them, and with whom he had contracted an intimacy on the passage. His friend very soon put an end to his doubts. 'He let me,' says Franklin, 'into Keith's character; told me there was not the least probability that he had written any letters for me; that no one who knew him had the smallest dependence on him; and he laughed at the idea of the Governor's giving me a letter of credit, having, as he said, no credit to give.'

Thus thrown once more on his own means, our young adventurer found there was no resource for him but to endeavour to procure some employment at his trade in London. Accordingly, having applied to a Mr. Palmer, a printer of eminence in Bartholomew Close, his services were accepted, and he remained there for nearly a year. During this time, although he was led into a good deal of idleness by the example of a friend, somewhat older than himself, he by no means forgot his old habits of reading and study. Having been employed in printing a second edition of Wollaston's *Religion of Nature*, his perusal of the work induced him to compose and publish a small pamphlet in refutation of some of the author's positions, which, he tells us, he did not afterwards look back upon as altogether a wise proceeding. He employed the greater part of his leisure more profitably in reading a great many works, which (circulating libraries, he remarks, not being then in use) he borrowed, on certain terms that were agreed upon between them, from a bookseller, whose shop was next door to his lodgings in Little Britain, and who had an immense collection of second-hand books. His pamphlet, however, was the means of making him known to a few of the literary characters then in London, among the rest to the noted Dr. Mandeville, author of the *Fable of the Bees;* and to Dr. Pemberton, Sir Isaac Newton's friend, who promised to give him an opportunity, some time or other, of seeing that great man; but this, he says, never happened. He also became acquainted about the same time with the famous collector and naturalist, Sir Hans Sloane, the founder of the British Museum, who had heard of some curiosities which Franklin had brought over from America; among these was a purse made of *asbestos*, which he purchased from him.

While with Mr. Palmer, and afterwards with Mr. Watts, near Lincoln's Inn Fields, he gave very striking evidence of those habits of temperance, self-command, industry, and frugality which distinguished him through after-life, and were undoubtedly the source of much of the success that attended his persevering efforts to raise himself from the humble condition in which he passed his earlier years. While Mr. Watts' other workmen spent a great part of every week's wages on beer, he drank only water, and found himself a good deal stronger, as well as much more clear-headed, on his light beverage than they on their strong potations. 'From my example,' says he, 'a great many of them left off their muddling breakfast of beer, bread, and cheese, finding they could with me be supplied from a neighbouring house with a large porringer of hot-water gruel, sprinkled with pepper, crumbled with bread, and a bit of butter in it, for the price of a pint of beer, viz. three halfpence. This was a more comfortable, as well as a cheaper breakfast, and kept their heads clearer. Those who continued sotting with their beer all day, were often, by not paying, out of credit at the alehouse, and used to make interest with me to get beer—*their light*, as they phrased it, *being out*. I watched the pay-table on Saturday night, and collected what I stood engaged for them, having to pay sometimes near thirty shillings a week on their accounts. This, and my being esteemed a pretty good *riggite*, that is, a jocular verbal satirist, supported my consequence in the society. My constant attendance (I never making a *St. Monday*) recommended me to the master; and my uncommon quickness at composing occasioned my being put upon works of despatch, which are generally better paid; so I went on now very agreeably.'

He spent about eighteen months altogether in London, during most part of which time he worked hard, he says, at his business, and spent but little upon himself except in seeing plays, and in books. At last his friend Mr. Denham, the gentleman with whom, as we mentioned before, he had got acquainted on his voyage to England, informed him he was going to return to Philadelphia to open a store, or mercantile establishment, there, and offered him the situation of his clerk at a salary of fifty pounds. The money was less than he was now making as a compositor; but he longed to see his native country again, and he accepted the proposal. Accordingly, they set sail together; and, after a long voyage, arrived in Philadelphia on the 11th of October 1726. Franklin was at this time only in his twenty-first year, and he mentions having formed, and committed to writing, while at sea, a plan for regulating the future conduct of his life. This unfortunately has been lost; but he tells us himself, that although conceived and determined upon when he was so young, it had yet 'been pretty faithfully adhered to quite through to old age.'

Mr. Denham had only begun business for a few months when he died; and Franklin was once more left upon the world. He now engaged again with his old master, Keimer, the printer, who had got a better house, and plenty of new types, though he was still as ignorant of his business as he was at the time of Franklin's former connection with him. While in this situation, Franklin got acquainted with several persons, like himself, fond of literary pursuits; and as the men never worked on Saturday, that being Keimer's self-appointed Sabbath, he had the whole day for reading.[1] He also showed his ingenuity, and the fertility of his resources, on various occasions. They wanted some new types, which, there being no letter foundry in America, were only to be procured from England; but Franklin, having seen types cast in London, though he had paid no particular attention to the process, contrived a mould, made use of the letters they had as puncheons, struck the matrices in lead, and thus supplied, as he tells us, in a pretty tolerable way, all deficiencies. 'I also,' he adds, 'engraved several things on occasion; made the ink; I was warehouseman; and, in short, quite a *factotum*.'

He did not, however, remain long with Keimer, who had engaged him only that he might have his other workmen

---

[1] Keimer had peculiar notions upon religious observances, and amongst other things fancied it a Christian duty to observe the Sabbath on the last day of the week.

taught through his means; and, accordingly, when this object was in some sort attained, contrived to pick a quarrel with him, which produced an immediate separation. He then entered into an agreement with one of his fellow-workmen, of the name of Meredith, whose friends were possessed of money, to begin business in Philadelphia in company with him, the understanding being that Franklin's skill should be placed against the capital to be supplied by Meredith. While he and his friend, however, were secretly preparing to put their plan in execution, he was induced to return for a few months to Keimer, on his earnest invitation, to enable him to perform a contract for the printing of some paper-money for the State of New Jersey, which required a variety of cuts and types that nobody else in the place could supply; and the two having gone together to Burlington to superintend this business, Franklin was fortunate enough, during the three months he remained in that city, to acquire, by his agreeable manners and intelligent conversation, the friendship of several of the principal inhabitants, with whom his employment brought him into connection. Among these he mentions particularly Isaac Decow, the surveyor-general. 'He was,' says Franklin, 'a shrewd, sagacious old man, who told me that he began for himself, when young, by wheeling clay for the brickmakers, learned to write after he was of age, carried the chain for surveyors, who taught him surveying, and he had now by his industry acquired a good estate; and, said he, I foresee that you will soon work this man (Keimer) out of his business, and make a fortune in it at Philadelphia. He had then not the least intimation of my intention to set up there or anywhere.'

Soon after he returned to Philadelphia the types that had

been sent for from London arrived; and, settling with Keimer, he and his partner took a house, and commenced business. 'We had scarce opened our letters,' says he, 'and put our press in order, before George House, an acquaintance of mine, brought a countryman to us, whom he had met in the street, inquiring for a printer. All our cash was now expended in the variety of particulars we had been obliged to procure, and this countryman's five shillings, being our first-fruits, and coming so seasonably, gave me more pleasure than any crown I have since earned; and, from the gratitude I felt towards House, has made me often more ready than perhaps I otherwise should have been, to assist young beginners.' He had, in the autumn of the preceding year, suggested to a number of his acquaintances a scheme for forming themselves into a club for mutual improvement; and they had accordingly been in the habit of meeting every Friday evening under the name of the Junto. All the members of this association exerted themselves in procuring business for him; and one of them, named Breinthal, obtained from the Quakers the printing of forty sheets of a history of that sect of religionists, then preparing at the expense of the body. 'Upon these,' says Franklin, 'we worked exceeding hard, for the price was low. It was a folio. I composed a sheet a day, and Meredith worked it off at press. It was often eleven at night, and sometimes later, before I had finished my distribution for the next day's work; for the little jobs sent in by our other friends, now and then, put us back. But so determined was I to continue doing a sheet a day of the folio, that one night, when, having imposed my forms, I thought my day's work over, one of them by accident was broken, and two pages (the half of the day's work) reduced to *pie*, I immediately distributed and composed it over again before I went to bed; and this industry, visible to our neighbours, began to give us character and credit.' The consequence was that business, and even offers of credit, came to them from all hands.

They soon found themselves in a condition to think of establishing a newspaper; but Franklin having inadvertently mentioned this scheme to a person who came to him wanting employment, that individual carried the secret to their old master, Keimer, with whom he, as well as themselves, had formerly worked; and he immediately determined to anticipate them by issuing proposals for a paper of his own. The manner in which Franklin met and defeated this treachery is exceedingly characteristic. There was another paper published in the place, which had been in existence for some years; but it was altogether a wretched affair, and owed what success it had merely to the absence of all competition. For this print, however, Franklin, not being able to commence his own paper immediately, in conjunction with a friend, set about writing a series of amusing communications under the title of the Busy Body, which the publisher printed, of course, very gladly. 'By this means,' says he, 'the attention of the public was fixed on that paper; and Keimer's proposals, which we burlesqued and ridiculed, were disregarded. He began his paper, however; and before carrying it on three-quarters of a year, with at most only ninety subscribers, he offered it me for a trifle; and I, having been ready some time to go on with it, took it in hand directly, and it proved in a few years extremely profitable to me.' The paper, indeed, had no sooner got into Franklin's hands than its success equalled his most sanguine expectations. Some observations which he wrote and printed in it on a colonial subject, then much talked of, excited so much attention among the leading people of the place, that it obtained the proprietors many friends in the House of Assembly, and they were, on the first opportunity, appointed printers to the House. Fortunately, too, certain events occurred about this time which ended in the dissolution of Franklin's connection with Meredith, who was an idle, drunken fellow, and had all along been a mere incumbrance upon the concern. His father failing to advance the capital which had been agreed upon, when payment was demanded at the usual time by their paper merchant and other creditors, he proposed to Franklin to relinquish the partnership and leave the whole in his hands, if the latter would take upon him the debts of the company, return to his father what he had advanced on their commencing business, pay his little personal debts, and give him thirty pounds and a new saddle. By the kindness of two friends, who, unknown to each other, came forward unasked to tender their assistance, Franklin was enabled to accept of this proposal; and thus, about the year 1729, when he was yet only in the twenty-fourth year of his age, he found himself, after all his disappointments and vicissitudes, with nothing, indeed, to depend upon but his own skill and industry for gaining a livelihood, and from extricating himself from debt, but yet in one sense fairly established in life, and with at least a prospect of well-doing before him.

Having followed his course thus far with so minute an observance of the several steps by which he arrived at the point to which we have now brought him, we shall not attempt to pursue the remainder of his career with the same particularity. His subsequent efforts in the pursuit of fortune and independence were, as is well known, eminently successful; and we find in his whole history, even to its close, a display of the same spirit of intelligence and love of knowledge, and the same active, self-denying, and intrepid virtues, which so greatly distinguished its commencement. The publication of a pamphlet, soon after Meredith had left him, in recommendation of a paper currency, a subject then much debated in the province, obtained him such popularity that he was employed by the Government in printing the notes after they had resolved upon issuing them. Other profitable business of the same kind succeeded. He then opened a stationer's shop, began gradually to pay off his debts, and soon after married. By this time his old rival, Keimer, had gone to ruin; and he was (with the exception of an old man, who was rich, and did not care about business) the only printer in the place. We now find him taking a leading part as a citizen. He established a circulating library, the first ever known in America, which, although it commenced with only fifty subscribers, became in course of time a large and valuable collection, the proprietors of which were eventually incorporated by royal charter. While yet in its infancy, however, it afforded its founder facilities of improvement of which he did not fail to avail himself, setting apart, as he tells us, an hour or two every day for study, which was the only amusement he allowed himself. In 1732, he first published his celebrated Almanac, under the name of *Richard Saunders*, but which was commonly known by the name of Poor Richard's Almanack. He continued this publication annually for twenty-five years. The proverbs and pithy sentences scattered up and down in the different numbers of it, were afterwards thrown together into a connected discourse under the title of *The Way to Wealth*, a production which has become so extensively popular that many of our readers are probably familiar with it.

We shall quote, in his own words, the account he gives us of the manner in which he pursued one branch of his studies :—

'I had begun,' says he, 'in 1733 to study languages. I soon made myself so much a master of the French, as to be able to read the books in that language with ease. I then undertook the Italian. An acquaintance, who was also learning it, used often to tempt me to play chess with him. Finding this took up too much of the time I had to spare for study, I at length refused to play any more, unless on this condition, that the victor in every game should have a right to impose a task, either of parts of the grammar to be got by heart, or in translations, etc., which tasks the vanquished was to perform upon honour before our next meeting. As we played pretty equally, we thus beat one another into that language. I afterwards, with a little painstaking, acquired as much of the Spanish as to read their books also. I have already mentioned that I had had only one year's instruction in a Latin school, and that when very young, after which I neglected that language entirely. But when I had attained an acquaintance with the French, Italian, and Spanish, I was surprised to find, on looking over a Latin Testament, that I understood more of that language than I had imagined, which encouraged me to apply myself again to the study of it; and I met with the more success, as those preceding languages had greatly smoothed my way.'

In 1736 he was chosen clerk of the General Assembly, and being soon after appointed deputy-postmaster for the

State, he turned his thoughts to public affairs, beginning, however, as he says, with small matters. He first occupied himself in improving the city watch; then suggested and promoted the establishment of a fire insurance company; and afterwards exerted himself in organizing a philosophical society, an academy for the education of youth, and a militia for the defence of the province. In short, every part of the civil government, as he tells us, and almost at the same time, imposed some duty upon him. 'The Governor,' he says, 'put me into the commission of the peace; the corporation of the city chose me one of the common council, and soon after alderman; and the citizens at large elected me a burgess to represent them in assembly. This latter station was the more agreeable to me, as I grew at length tired with sitting there to hear the debates, in which, as clerk, I could take no part, and which were often so uninteresting that I was induced to amuse myself with making magic squares or circles, or anything to avoid weariness; and I conceived my becoming a member would enlarge my power of doing good. I would not, however, insinuate that my ambition was not flattered by all these promotions,—it certainly was; for considering my low beginning, they were great things to me; and they were still more pleasing as being so many spontaneous testimonies of the public good opinion, and by me entirely unsolicited.'

It is time, however, that we should introduce this extraordinary man to our readers in a new character. A much more important part in civil affairs than any he had yet acted was in reserve for him. He lived to attract to himself on the theatre of politics, the eyes not of his own countrymen only, but of the whole civilised world; and to be a principal agent in the production of events as mighty in themselves, and as pregnant with mighty consequences, as any belonging to modern history. But our immediate object is to exhibit a portrait of the diligent student, and of the acute and patient philosopher. We have now to speak of Franklin's famous electrical discoveries. Of these discoveries we cannot, of course, here attempt to give anything more than a very general account.

The term electricity is derived from *electron*, the Greek name for amber, which was known, even in ancient times, to be capable of acquiring, by being rubbed, the curious property of attracting very light bodies, such as small bits of paper, when brought near to them. This virtue was thought to be peculiar to the substance in question, and one or two others, down to the close of the sixteenth century, when our ingenious and philosophic countryman, William Gilbert, a physician of London, announced for the first time, in his Latin treatise on the magnet, that it belonged equally to the diamond and many other precious stones; to glass, sulphur, sealing wax, rosin, and a variety of other substances. It is from this period that we are to date the birth of the science of Electricity, which, however, continued in its infancy for above a century, and could hardly, indeed, be said to consist of anything more than a collection of unsystematized and ill-understood facts until it attracted the attention of Franklin.

Among the facts, however, that had been discovered in this interval, the following were the most important. In the first place, the list of the substances capable of being excited by friction to a manifestation of electric virtue was considerably extended. It was also found that the bodies which had been attracted by the excited substance were immediately after as forcibly repelled by it, and could not be again attracted until they had touched a third body. Other phenomena, too, besides those of attraction and repulsion, were found to take place when the body excited was one of sufficient magnitude. If any other body not capable of being excited, such as the human hand or a rod of metal, was presented to it, a slight sound would be produced, which, if the experiment was performed in a dark room, would be accompanied with a momentary light. Lastly, it was discovered that the electric virtue might be imparted to bodies not capable of being themselves excited ; by making such a body, when insulated, that is to say separated from all other bodies of the same class by the intervention of one capable of excitation, act either as the rubber of the excited body, or as the drawer of a succession of sparks from it, in the manner that has just been described. It was said, in either of these cases, to be *electrified ;* and it was found that if it was touched, or even closely approached, when in this state, by any other body, in like manner incapable of being excited by friction, a pretty loud report would take place, accompanied, if either body was susceptible of feeling, with a slight sensation of pain at the point of contact, and which would instantly restore the electrified body to its usual and natural condition.

In consequence of its thus appearing that all those bodies, and only those, which could not be themselves excited, might in this manner have electricity, as it were, transferred to them, they were designated *conductors*, as well as *non-electrics ;* while all *electrics*, on the other hand, were also called *non-conductors*. It is proper, however, that the reader should be aware, that of the various substances in nature, none, strictly speaking, belong exclusively to either of these classes; the truth being merely, that different bodies admit the passage of the electric influence with extremely different degrees of facility, and that those which transmit it readily are called conductors,—the metals, and fluids, and living animals particularly belonging to this class; while such as resist its passage, or permit it only with extreme reluctance,—among which are amber, sulphur, wax, glass, and silk,—are described by the opposite denomination.

The beginning of the year 1746 is memorable in the annals of electricity for the accidental discovery of the possibility of accumulating large quantities of the electric fluid by means of what was called the Leyden jar, or phial. M. Cuneus, of that city, happened one day, while repeating some experiments which had been originally suggested by M. von Kleist, Dean of the Cathedral in Camin, to hold in one hand a glass vessel, nearly full of water, into which he had been sending a charge from an electrical machine, by means of a wire dipped into it, and communicating with the prime conductor, or insulated non-electric, exposed in the manner we have already mentioned to the action of the excited cylinder. He was greatly surprised, upon applying his other hand to disengage the wire from the conductor, when he thought that the water had acquired as much electricity as the machine could give it, by receiving a sudden shock in his arms and breast, much more severe than anything of the kind he had previously encountered in the course of his experiments. The same thing, it was found, took place when the glass was covered, both within and without, with any other conductors than the water and the human hand, which had been used in this instance; as, for example, when it was coated on both sides with tinfoil, in such a manner, however, that the two coatings were completely separated from each other, by a space around the lip of the vessel being left uncovered. Whenever a communication was formed by the interposition of a conducting medium between the inside and outside coating, an instant and loud explosion took place, accompanied with a flash of light, and the sensation of a sharp blow, if the conductor employed was any part of the human body.

The first announcement of the wonders of the Leyden phial excited the curiosity of all Europe. The accounts given of the electric shock by those who first experienced it are perfectly ludicrous, and well illustrate how strangely the imagination is acted upon by surprise and terror, when novel or unexpected results suddenly come upon it.

From the original accounts, as Dr. Priestley observes, could we not have repeated the experiment, we should have formed a very different idea of the electric shock to what it really is, even when given in greater strength than it could have been by those early experimenters. It was this experiment, however, that first made electricity a subject of general curiosity. Everybody was eager, notwithstanding the alarming reports that were spread of it, to feel the new sensation; and in the same year in which the experiment was first made at Leyden, numbers of persons, in almost every country in Europe, obtained a livelihood by going about and showing it.

The particulars, then, that we have enumerated may be said to have constituted the whole of the science of Electricity, in the shape in which it first presented itself to the notice of Dr. Franklin. In the way in which we have stated them, they are little more, the reader will observe, than a mass of seemingly unconnected facts, having, at first sight, no semblance whatever of being the results of a common principle, or of being reducible to any general and comprehensive system. It is true that a theory, that of M. Dufay, had been formed before this time to account for many of them, and also for others that we have not mentioned; but it does not appear that Franklin ever heard of it until he had formed his own, which is, at all events, entirely different; so that it is unnecessary for us to take it at all into account. We shall form a fair estimate of the amount and merits of Franklin's discoveries, by considering the facts we have mentioned as really constituting the science in the state in which he found it.

It was in the year 1746, as he tells us himself in the narrative of his life, that, being in Boston, he met with a Dr. Spence, who had lately arrived from Scotland, and who showed him some electrical experiments. They were imperfectly performed, as the doctor was not very expert; 'but being,' says Franklin, 'on a subject quite new to me, they equally surprised and pleased me. Soon after my return to Philadelphia, our Library Company received from Mr. Peter Collinson, F.R.S. of London, a present of a glass tube, with some account of the use of it in making such experiments. I eagerly seized the opportunity of repeating what I had seen at Boston, and, by much practice, acquired great readiness in performing those also which we had an account of from England, adding a number of new ones. I say much practice, for my house was continually full for some time with persons who came to see these new wonders. To divide a little this incumbrance among my friends, I caused a number of similar tubes to be blown in our glasshouse, with which they furnished themselves, so that we had at length several performers.' The newly-discovered and extraordinary phenomena exhibited by the Leyden phial, of course, very early engaged his attention in pursuing these interesting experiments; and his inquisitive mind immediately set itself to work to find out the reason of such strange effects, which still astonished and perplexed the ablest philosophers of Europe. Out of his speculations arose the ingenious and beautiful theory of the action of the electric influence which is known by his name, and which was at that time received by the greater number of philosophers as the best, because the simplest and most complete, demonstration of the phenomena that had until then been given to the world.

Dr. Franklin's earliest inquiries were directed to ascertain the *source* of the electricity which friction had the effect of at least rendering manifest in the glass cylinder, or other electric. The question was whether this virtue was created by the friction in the electric, or only thereby communicated to it from other bodies. In order to determine this point, he resorted to the very simple experiment of endeavouring to electrify himself; that is to say, having insulated himself and excited the cylinder by rubbing it with his hand, he then drew off its electricity from it in the usual manner into his own body. But he found that he was not thereby electrified at all, as he would have been by doing the same thing had the friction been applied by another person. No spark could be obtained from him, after the operation, by the presentment of a conductor; nor did he exhibit on such bodies as were brought near him any of the other usual evidences of being charged with electricity.

If the electricity had been created in the electric by the friction, it was impossible to conceive why the person who drew it off should not have been electrified in this case, just as he would have been had another person acted as the rubber. The result evidently indicated that the friction had effected a change upon the person who had performed that operation, as well as upon the cylinder, since it had rendered him incapable of being electrified by a process by which, in other circumstances, he would have been so. It was plain, in short, that the electricity had passed, in the first instance, out of his body into the cylinder; which, therefore, in communicating it to him in the second instance, only gave him back what it had received, and, instead of electrifying him, merely restored him to his usual state—to that in which he had been before the experiment was begun.

This accordingly was the conclusion to which Franklin came; but, to confirm it, he next insulated two individuals, one of whom he made to rub the cylinder, while the other drew the electricity from it. In this case, it was not the latter merely that was affected; both were electrified. The one had given out as much electricity to the cylinder in rubbing it, as the other had drawn from it. To prove this still further, he made them touch one another, when both were instantly restored to their usual state, the redundant electricity thrown off by the one exactly making up the deficiency of the other. The spark produced by their contact was also, as was to have been expected, greater than that which took place when either of them was touched by any third person who had not been electrified.

Proceeding upon the inferences which these results seemed so evidently to indicate, Franklin constructed the general outlines of his theory. Every body in nature he considered to have its natural quantity of electricity, which may, however, be either diminished, by part of it being given out to another body, as that of the rubber, in the operation of the electrical machine, is given out to the cylinder; or increased, as when the body is made to receive the electricity from the cylinder.

In the one case he regarded the body as *negatively*, in the other as *positively*, electrified. In the one case it had less, in the other more, than its natural quantity of electricity; in either, therefore, supposing it to be composed of electricity and common matter, the usual equilibrium or balance between its two constituent ingredients was, for the time, upset or destroyed.

But how should this produce the different effects which are observed to result from the action of electrified bodies? How is the mere circumstance of the overthrow of the customary equilibrium, between the electricity and the matter of a body, to be made to account for its attraction and repulsion of other bodies, and for the extraordinary phenomena presented by the Leyden phial? The Franklinian theory answers these questions with great ease and completeness.

The fundamental law of the electric fluid, according to this theory, is that its particles attract matter, and repel one another. To this we must add a similar law with regard to the particles of matter, namely, that they repel each other, as well as attract electricity. This latter consideration was somewhat unaccountably overlooked by Franklin, but was afterwards introduced by Mr. Æpinus, of St. Petersburg, and our celebrated countryman, the late Mr. Cavendish, in their more elaborate expositions of his theory of the electrical action. Let us now apply these two simple principles to the explanation of the facts we have already mentioned.

In the first place, when two bodies are in their ordinary or natural state, the quantity of matter is an exact balance for the quantity of electricity in each, and there is accordingly no tendency of the fluid to escape; no spark will take place between two such bodies when they are brought into contact.

Nor will they either attract or repel each other, because the attractive and repulsive forces operating between them are exactly balanced, the two attractions of the electricity in the first for the matter in the second, and of the electricity in the second for the matter in the first, being opposed by the two repulsions of the electricity in the first for the electricity in the second, and of the matter in the first for the matter in the second. They, therefore, produce no effect upon each other whatever.

But let us next suppose that one of the bodies is an electric which has been excited in the usual way by friction—a stick of wax, or a glass cylinder, for example, which has been rubbed with the hand, or a piece of dry silk. In this case, the body in question has received an addition to its natural quantity of electricity, which addition, accordingly, it will most readily part with whenever it is brought into contact with a conductor. But this is not all. Let us see how it will act, according to the law that has been stated, upon the other body, which we shall suppose to be in its natural state, when they are brought near each other. First, from the repulsive tendency of the electric particles, the extra electricity in the excited body will drive away a portion of the electricity of the other from its nearest end, which will thus become negatively electrified, or will consist of more matter than is necessary to balance its electricity. In this state of things, what are the attractive and repulsive forces operating between the two bodies, the one, be it remembered, having an excess of electricity, and the other an excess of matter? There are, in fact, five attractive forces opposed by only four repulsive— the former being those of the matter in the first body for the electricity in the second, of the balanced electricity in the first for the balanced matter in the second, of the same for the extra matter in the second, together with the two of the extra electricity in the first for the same two quantities of matter ; and the latter being those of the matter in the first for the balanced matter in the second, of the same for the extra matter in the second, together with those of the electricity in the second both for the balanced and the extra electricity in the first. The two bodies, therefore, ought to meet, as we find they actually do. But no sooner do they meet than the extra electricity of the first, attracted by the matter of the second, flows over partly to it ; and both bodies become positively electrified—that is to say, each contains a quantity of electricity beyond that which its matter is capable of balancing. It will be found, upon examination, that we have now four powers of attraction opposed by five of repulsion—the former being those of the matter in each body for the two electricities in the other ; the latter, those exerted by each of the electricities in the one against both the electricities of the other, together with that of the matter in the one for the matter in the other. The bodies now accordingly should repel each other, just as we find to be the fact. Of course, the same reasoning applies to the case of a neutral body, and any other containing a superabundance of electricity, whether it be an electric or no, and in whatever way its electricity may have been communicated to it. We may add, that there is no case of attraction or repulsion between two bodies, in which the results indicated by the theory do not coincide with those of observation as exactly as in this.

We now come to the phenomena of the Leyden phial. The two bodies upon which we are here to fix our attention are the interior and exterior coatings, which, before the process of charging has commenced, are of course in their natural state, each having exactly that quantity of electricity which its matter is able to balance, and neither therefore exerting any effect whatever upon the other. But no sooner has the interior coating received an additional portion of electricity from the prime conductor, with which the reader will remember it is in communication, than, being now positively electrified, it repels a corresponding portion of its electricity from the exterior coating, which therefore becomes negatively electrified. As the operation goes on, both these effects increase, till at last the superabundance of electricity in the one surface, and its deficiency in the other, reach the limit to which it is wished to carry them. All this while, it will be remarked, the former is prevented from giving out its superfluity to the latter by the interposition of the glass, which is a non-conductor, and the uncovered space which had been left on both sides around the lip of the vessel. If the charge were made too high, however, even these obstacles would be overcome, and the unbalanced electricity of the interior coating, finding no easier vent, would at last rush through the glass to the unsaturated matter on its opposite surface, probably shattering it to pieces in its progress. But, to effect a discharge in the usual manner, a communication must be established by means of a good conductor between the two surfaces, before this extreme limit be reached. If either a rod of metal, for example, or the human body, be employed for this purpose, the fluid from the interior coating will instantly rush along the road made for it, occasioning a pretty loud report, and, in the latter case, a severe shock, by the rapidity of its passage.

Both coatings will, in consequence, be immediately restored to their natural state.

That this is the true explanation of the matter, Franklin further demonstrated by a variety of ingenious experiments. In the first place, he found that, if the outer coating was cut off, by being insulated from every conducting body, the inner coating could not be charged; the electricity in the outer coating had here no means of escape, and it was consequently impossible to produce in that coating the requisite negative electricity. On the other hand, if a good conductor was brought within the striking distance from the outside coating, while the process of charging was going on, the expelled fluid might be seen passing away towards it in sparks, in proportion as more was sent from the prime conductor into the inside of the vessel. He observed also that, when a phial was charged, a cork ball, suspended on silk, would be attracted by the one coating when it had been repelled by the other—an additional indication and proof of their opposite states of electricity, as might be easily shown by an analysis of the attractive and repulsive forces operating between the two bodies in each case.

But Franklin did not rest contented with ascertaining the principle of the Leyden phial. He made also a very happy application of this principle, which afforded a still more wonderful manifestation than had yet been obtained of the powers of accumulated electricity. Considering the waste that took place, in the common experiment, of the fluid expelled, during the process of charging, from the exterior coating, he conceived the idea of employing it to charge the inner surface of a second jar, which he effected, of course, by the simple expedient of drawing it off by means of a metal rod communicating with that surface. The electricity expelled from the outside of this second jar was conveyed, in like manner, into the inside of a third; and, in this way, a great number of jars were charged with the same facility as a single one. Then, having connected all the inside coatings with one conductor, and all the outside coatings with another, he had merely to bring these two general conductors into contact or communication, in order to discharge the whole accumulation at once. This contrivance he called an *Electrical Battery*.

The general sketch we have thus given will put the reader in possession, at least, of the great outlines of the Franklinian theory of electricity, undoubtedly one of the most beautiful generalizations to be found in the whole compass of science. By the aid of what we may call a single principle, since the law with regard to the electric fluid and common matter is exactly the same, it explains satisfactorily not only all the facts connected with this interesting subject which were known when it was first proposed, but all those that have been since discovered, diffusing order and light throughout what seemed before little better than a chaos of unintelligible contradictions. We must now, however, turn to a very brilliant discovery of this illustrious philosopher, the reality of which does not depend upon the truth or falsehood of any theory.

Franklin was by no means the first person to whom the idea had suggested itself of a similarity between electricity and lightning. Not to mention many other names which might be quoted, the Abbé Nollet had, before him, not only intimated his suspicion that thunder might be in the hands of Nature what electricity is in ours, but stated a variety of reasons on which he rested his conjecture. It is to Franklin alone, however, that the glory belongs of both pointing out the true method of verifying this conjecture, and of actually establishing the perfect identity of the two powers in question. 'It has, indeed, been of late the fashion,' says the editor of the first account of his electrical experiments, published at London in 1751, 'to ascribe every grand or unusual operation of nature, such as lightning and earthquakes, to electricity; not, as one would imagine from the manner of reasoning on these occasions, that the authors of these schemes have discovered any connection betwixt the cause and effect, or saw in what manner they were related; but, as it would seem, merely because they were unacquainted with any other agent, of which it could not positively be said the connection was impossible.' Franklin transformed what had been little more than a figure of rhetoric into a most important scientific fact.

In a paper, dated November 7, 1749, he enumerates all the known points of resemblance between lightning and electricity. In the first place, he remarks, it is no wonder that the effects of the one should be so much greater than those of the other; for if two gun-barrels electrified will strike at two inches' distance, and make a loud report, at how great a distance will ten thousand acres of electrified cloud strike, and give its fire; and how loud must be that crack! He then notices the crooked and waving course both of the flash of lightning and, in some cases, of the electric sparks; the tendency of lightning, like electricity, to take the readiest and best conductor; the fact that lightning, as well as electricity, dissolves metals, burns some bodies, rends others, strikes people blind, destroys animal life, reverses the poles of magnets, etc.

He had known for some time the extraordinary power of pointed bodies, both in drawing and in throwing off the electric fire. The true explanation of this fact did not occur to him; but it is a direct consequence of the fundamental principle of his own theory, according to which the repulsive tendency of the particles of electricity towards each other, occasioning the fluid to retire, in every case, from the interior to the surface of bodies, drives it with especial force towards points and other prominences, and thus favours its escape through such outlets; while, on the other hand, the more concentrated attraction which the matter of a pointed body, as compared with that of a blunt one, exerts upon the electricity to which it is presented, brings it down into its new channel in a denser stream. In possession, however, of the fact, we find him concluding the paper we have mentioned as follows:—'The electric fluid is attracted by points. We do not know whether this property be in lightning; but since they agree in all the particulars in which we can already compare them, it is not improbable that they agree likewise in this. Let the experiment be made.'

Full of this idea, it was yet some time before he found what he conceived a favourable opportunity of trying its truth in the way he meditated. A spire was about to be erected in Philadelphia, which he thought would afford him facilities for the experiment; but his attention having been one day drawn by a kite which a boy was flying, it suddenly occurred to him that here was a method of reaching the clouds preferable to any other. Accordingly, he immediately took a large silk handkerchief and stretching it over two cross sticks, formed in this manner his simple apparatus for drawing down the lightning from its cloud. Soon after, seeing a thunderstorm approaching, he took a walk into a field in the neighbourhood of the city, in which there was a shed,—communicating his intentions, however, to no one but his son, whom he took with him, to assist him in raising the kite: this was in June 1752.

The kite being raised, he fastened a key to the lower extremity of the hempen string, and then insulating it by attaching it to a post by means of silk, he placed himself under the shed, and waited the result. For some time no signs of electricity appeared. A cloud, apparently charged with lightning, had even passed over them without producing any effect. At length, however, just as Franklin was beginning to despair, he observed some loose threads of the hempen string rise and stand erect, exactly as if they had been repelled from each other by being charged with electricity. He immediately presented his knuckle to the key, and, to his inexpressible delight, drew from it the well-known electrical spark. It is said that his emotion was so great at this completion of a discovery which was to make his name immortal, that he heaved a deep sigh, and felt that he could that moment have willingly died. As the rain increased, the cord became a better conductor, and the key gave out its electricity copiously. Had the hemp been thoroughly wet, the bold experimenter might, as he was contented to do, have paid for his discovery with his life.

He afterwards brought down the lightning into his house, by means of an insulated iron rod, and performed with it, at his leisure, all the experiments that could be performed with electricity. But he did not stop here. His active and practical mind was not satisfied even with the splendid discovery, until he had turned it to a useful end. It suggested to him, as is well known, the idea of a method of preserving buildings from lightning, which is extremely simple and cheap, as well as effectual, consisting, as it does, in nothing more than attaching to the building a pointed metallic rod, rising higher than any part of it, and communicating at the lower end with the ground. This rod the lightning is sure to seize upon in preference to any part of the building; by which means it is conducted to the earth, and prevented from doing any injury. There was always a strong tendency in Franklin's philosophy to these practical applications. The lightning-rod was probably the result of some of the amusing experiments with which Franklin was, at the commencement of his electrical investigations, accustomed to employ his own leisure, and afford pleasure to his friends. In one of his letters to Mr. Collinson, dated so early as 1748, we find him expressing himself in the following strain, in reference to his electrical experiments :—' Chagrined a little that we have hitherto been able to produce nothing in this way of use to mankind, and the hot weather coming on, when electrical experiments are not so agreeable, it is proposed to put an end to them for this season somewhat humorously, in a party of pleasure on the banks of *Skuylkill.* Spirits at the same time are to be fired by a spark sent from side to side through the river, without any other conductor than the water—an experiment which we have some time since performed to the amazement of many. A turkey is to be killed for dinner by the *electrical shock*, and roasted by the *electrical jack*, before a fire kindled by the *electrified bottle;* when the healths of all the famous electricians in England, Holland, France, and Germany are to be drunk in *electrified bumpers*, under the discharge of guns from the *electrical battery.*'

Franklin's electrical discoveries did not, on their first announcement, attract much attention in England; and, indeed, he had the mortification of learning that his paper on the similarity of lightning to electricity, when read by a friend to the Royal Society, had been only laughed at by that learned body. In France, however, the account that had been published in London of his experiments, fortunately fell into the hands of the celebrated naturalist Buffon, who was so much struck with it that he had it translated into French and printed at Paris. This made it immediately known to all Europe; and versions of it in various other modern languages soon appeared, as well as one in Latin. The theory propounded in it was at first violently opposed in France by the Abbé Nollet, who had one of his own to support, and, as Franklin tells us, could not at first believe that such a work came from America, but said it must have been fabricated by his enemies at Paris. The Abbé was eventually, however, deserted by all his partisans, and lived to see himself the last of his sect. In England, too, the Franklinian experiments gradually began to be more spoken of; and, at last, even the Royal Society was induced to resume the consideration of the papers that had formerly been read to them. One of their members verified the grand experiment of bringing down lightning from the clouds; and upon his reading to them an account of his success, 'they soon,' says Franklin, 'made me more than amends for the slight with which they had before treated me. Without my having made any application for that honour, they chose me a member; and voted that I should be excused the customary payments, which would have amounted to twenty-five guineas; and ever since have given me their *Transactions* gratis. They also presented me with the gold medal of Sir Godfrey Copley, for the year 1753, the delivery of which was accompanied with a very handsome speech of the President, Lord Macclesfield, wherein I was highly honoured.' Some years afterwards, when he was in this country with his son, the University of St. Andrews conferred upon him the degree of Doctor of Laws; and its example was followed by the Universities of Edinburgh and Oxford. He was also elected a member of many of the learned societies throughout Europe.

No philosopher of the age now stood on a prouder eminence than this extraordinary man, who had originally been one of the most obscure of the people, and had raised himself to all this distinction almost without the aid of any education but such as he had given himself. Who will say, after reading his story, that anything more is necessary for the attainment of knowledge than the determination to attain it?—that there is any other obstacle to even the highest degree of intellectual advancement which may not be overcome, except a man's own listlessness or indolence? The secret of this man's success in the cultivation of his mental powers was, that he was ever awake and active in that business; that he suffered no opportunity of forwarding it to escape him unimproved; that, however poor, he found at least a few pence, were it even by diminishing his scanty meals, to pay for the loan of the books he could not buy; that, however hard wrought, he found a few hours in the week, were it by sitting up half the night after toiling all the day, to read and study them. Others may not have his original powers of mind; but his industry, his perseverance, his self-command, are for the imitation of all; and though few may look forward to the rare fortune of achieving discoveries like his, all may derive both instruction and encouragement from his example. They who may never overtake the light, may at least follow its path, and guide their footsteps by its illumination.

Were we to pursue the remainder of Franklin's history, we should find the fame of the patriot vying with that of the philosopher in casting a splendour over it; and the originally poor and unknown tradesman standing before kings, associating as an equal with the most eminent statesmen of his time, and arranging along with them the wars and treaties of mighty nations. When the struggle of American independence commenced, he was sent as ambassador from the United States to the court of France, where he soon brought about an alliance between the two countries, which produced an immediate war between the latter and England. In 1783, he signed, on the part of the United States, the treaty of peace with England, which recognised their independence. Two years after he returned to his native country, where he was received with acclamation by his grateful and admiring fellow-citizens, and immediately elected President of the Supreme Executive Council. He closed his eventful and honourable life on the 17th of April 1790, in the eighty-fifth year of his age.[1]

---

[1] The most authentic edition of Franklin's memoirs is that entitled, *The Life of Benjamin Franklin*, written by himself. By John Bigelow. 3 vols. Lippincott & Co., Philadelphia.

JAMES BRINDLEY.

AMES BRINDLEY, the celebrated engineer, was entirely self-taught in even the rudiments of mechanical science,—although, unfortunately, we are not in possession of any very minute details of the manner in which his powerful genius first found its way to the knowledge of those laws of nature of which it afterwards made so many admirable applications. He was born at Tunsted, in the parish of Wormhill, Derbyshire, in the year 1716; and all we know of the first seventeen years of his life is that, his father having reduced himself to extreme poverty by his dissipated habits, he was allowed to grow up almost totally uneducated, and, from the time he was able to do anything, was employed in the ordinary descriptions of country labour. To the end of his life this great genius was barely able to read on any very pressing occasion; for, generally speaking, he would no more have thought of looking into a book for any information he wanted, than of seeking for it in the heart of a millstone; and his knowledge of the art of writing hardly extended farther than the accomplishment of signing his name. It is probable that, as he grew towards manhood, he began to feel himself created for higher things than driving a cart or following a plough; and we may even venture to conjecture, that the particular bias of his genius towards mechanical invention had already disclosed itself, when, at the age of seventeen, he bound himself apprentice to a person of the name of Bennet, a millwright, residing at Macclesfield, which was but a few miles from his native place. At all events, it is certain that he almost immediately displayed a wonderful natural aptitude for the profession he had chosen. 'In the early part of his apprenticeship,' says the writer of his life in the *Biographia Britannica*, who was supplied with the materials of his article by Mr. Henshall, Brindley's brother-in-law, 'he was frequently left by himself for whole weeks together to execute works concerning which his master had given him no previous instructions. These works, therefore, he finished in his own way; and Mr. Bennet was often astonished at the improvements his apprentice from time to time introduced into the millwright business, and earnestly questioned him from whom he had gained his knowledge. He had not been long at the trade before the millers, wherever he had been employed, always chose him again in preference to the master or any other workman; and before the expiration of his servitude, at which time Mr. Bennet, who was advanced in years, grew unable to work, Mr. Brindley, by his ingenuity and application, kept up the business with credit, and even supported the old man and his family in a comfortable manner.'

His master, indeed, does not appear to have been very capable of teaching him much of anything; and Brindley seems to have been left to pick up his knowledge of the business in the best way he could by his own observation and sagacity. Bennet having been employed on one occasion, we are told, to build the machinery of a paper mill, which he had never seen in his life, took a journey to a distant part of the country expressly for the purpose of inspecting one which might serve him for a model. However, he had made his observations, it would seem, to very little purpose; for, having returned home and fallen to work, he could make nothing of the business at all, and was only bewildering himself, when a stranger, who understood something of such matters, happening one day to see what he was about, felt no scruple in remarking in the neighbourhood that the man was only throwing away his employer's money. The reports which in consequence got abroad soon reached the ears of Brindley, who had been employed on the machinery under the directions of his master. Having probably of himself begun ere this to suspect that all was not right, his suspicions were only confirmed by what he heard; but, aware how unlikely it was that his master would be able to explain matters, or even to assist him in getting out of his difficulties, he did not apply to him. On the contrary, he said nothing to any one, but, waiting till the work of the week was over, set out by himself one Saturday evening to see the mill which his master had already visited. He accomplished his object, and was back to his work by Monday morning, having travelled the whole journey of fifty miles on foot. Perfectly master now of the construction of the mill, he found no difficulty in going on with his undertaking, and completed the machine, indeed, not only so as perfectly to satisfy the proprietor, but with several improvements on his model of his own contrivance.

After remaining some years with Bennet, he set up in business for himself. With the reputation he had already acquired, his entire devotion to his profession, and the wonderful talent for mechanical invention of which almost every piece of machinery he constructed gave evidence, he could not fail to succeed. But for some time, of course, he was known only in the neighbourhood of the place where he lived. His connections, however, gradually became more and more extensive, and at length he began to undertake engineering in all its branches. He distinguished himself greatly in 1752 by the erection of a water-engine for draining a coal mine at Clifton in Lancashire. The great difficulty in this case was to obtain a supply of water for working the engine; this he brought through a tunnel of six hundred yards in length, cut in the solid rock. It would appear, however, that his genius was not yet quite appreciated as it deserved to be, even by those who employed him. He was in some sort an intruder into his present profession, for which he had not been regularly educated; and it was natural enough that, before his great powers had had an opportunity of showing themselves, and commanding the universal admiration of those best qualified to judge of them, he should have been conceived by many to be rather a merely clever workman in a few particular departments, than one who could be safely entrusted with the entire management and superintendence of a complicated design. In 1755 it was determined to erect a new silk-mill at Congleton, in Cheshire; and, another person having been appointed to preside over the execution of the work, and to arrange the more intricate combinations, Brindley was engaged to fabricate the larger wheels and other coarser parts of the apparatus. It soon became manifest, however, in this instance, that the superintendent was unfit for his office, and the proprietors were obliged to apply to Brindley to remedy several blunders into which he had fallen, and give his advice as to how the work should be proceeded in. Still they did not deem it proper to dismiss their incapable projector, but, the pressing difficulty overcome, would have had him by whose ingenuity they had been enabled to get over it to return to his subordinate place and work under the directions of the same superior. This Brindley positively refused to do. He told them he was ready, if they would merely let him know what they wished the machine to perform, to apply his best endeavours to make it answer that purpose, and that he had no doubt he should succeed, but he would not submit to be superintended by a person whom he had discovered to be quite ignorant of the business he professed. This at once brought about a proper arrangement of matters. Brindley's services could not be dispensed with; those of the pretender who had been set over him might be so without much disadvantage. The entire management of the work, therefore, was forthwith confided to the former, who completed it with his usual ability in a superior manner. He not only made important improvements, indeed, in many parts of the machine itself, but even in the mode of preparing the separate pieces of which it was to be composed. His ever active genius was constantly displaying itself by the invention of the most beautiful and economical simplifications. One of these was a method which he contrived for cutting all his tooth and pinion wheels by machinery, instead of having them done by the hand, as they always till then had been. This invention enabled him to finish as much of that sort of work in one day as had formerly been accomplished in fourteen.

But the character of this man's mind was comprehensiveness and grandeur of conception; and he had not yet found any adequate field for the display of his vast ideas and almost inexhaustible powers of execution. Happily, however, this was at last afforded him, by the commencement of a series of undertakings in this country, which deservedly rank among the achievements of modern enterprise and mechanical skill, and which were destined, within no long period, to change the whole aspect of the internal commerce of the island.

Artificial water roads, or *canals*, were well known to the ancients. Without transcribing all the learning that has been collected upon the subject, and may be found in any of the common treatises, we may merely state that the Egyptians had early effected a junction by this means between the Red Sea and the Mediterranean; that both the Greeks and the Romans attempted to cut a canal across the Isthmus of Corinth; and the latter people actually cut one in Britain from the neighbourhood of Peterborough to that of Lincoln, some traces of which are still discernible. Canal navigation is also of considerable antiquity in China. The greatest work of this description in the world is the Imperial Canal of that country, which is two hundred feet broad, and, commencing at Pekin, extends southward to the distance of about nine hundred miles. It is supposed to have been constructed about eight centuries ago; but there are a great many smaller works of the same kind in the country, many of which are undoubtedly much older. The Chinese are unacquainted, as were also the ancients, with the contrivance called a lock, by means of which different levels are connected in many of our modern European canals, and which, as probably all our readers know, is merely a small intermediate space, in which the water can be kept at the same elevation as either part of the channel, into which the boat is admitted by the opening of one floodgate, and from which it is let out by the opening of another, after the former has been shut;—the purpose being thus attained of floating it onwards, without any greater waste of water than the quantity required to alter the level of the enclosed space. When locks are not employed, the canal must be either of uniform level throughout, or it must consist of a succession of completely separated portions of water-way, from the one to the other of which the boat is carried on an inclined plane, or by some other mechanical contrivance.

Canals have also been long in use in several of the countries of modern Europe, particularly in the Netherlands and in France. In the former, indeed, they constitute the principal means of communication between one place and another, whether for commercial or other purposes. In France, the canals of Burgundy, of Briare, of Orleans, and of Languedoc, all contribute important facilities to the commerce of the country. The last-mentioned, which unites the Mediterranean to the Atlantic, is sixty feet broad and one hundred and fifty miles in length. It was finished in 1681, having employed twelve thousand men for fifteen years, and cost twelve hundred thousand pounds sterling.

It is remarkable that, with these examples before her, England was so late in availing herself of the advantages of canal navigation. The subject, however, had not been altogether unthought of. As early as the reign of Charles the Second, a scheme was in agitation for cutting a canal (which has since been made) between the Forth and the

Clyde, in the northern part of the kingdom; but the idea was abandoned, from the difficulty of procuring the requisite funds. A very general impression, too, seems to have been felt, in the earlier part of the last century, as to the desirableness of effecting a canal navigation between the central English counties and either the metropolis or the eastern coast.

The first modern canal actually executed in England was not begun till the year 1755. It was the result of a sudden thought on the part of its undertakers, nothing of the kind having been contemplated by them when they commenced the operations which led to it. They had obtained an Act of Parliament for rendering navigable the Sankey brook, in Lancashire, which flows into the river Mersey, from the neighbourhood of the now flourishing town of St. Helens, through a district abounding in valuable beds of coal. Upon surveying the ground, however, with more care, it was considered better to leave the natural course of the stream altogether, and to carry the intended navigation along a new line; in other words, to cut a canal. The work was accordingly commenced; and the powers of the projectors having been enlarged by a second Act of Parliament, the canal was eventually extended to the length of about twelve miles. It turned out both a highly successful speculation for the proprietors and a valuable public accommodation.

It is probable that the Sankey canal, although it did not give birth to the first idea of the great work we are now about to describe, had at least the honour of prompting the first decided step towards its execution. Francis, Duke of Bridgewater, who, while yet much under age, had succeeded, in the year 1748, by the death of his elder brothers, to the family estates, and the title, which had been first borne by his father, had a property at Worsley, about seven miles west from Manchester, extremely rich in coal mines, which, however, had hitherto been unproductive, owing to the want of any sufficiently economical means of transport. The object of supplying this defect had for some time strongly engaged the attention of the young duke, as it had indeed done that of his father, who, in the year 1732, had obtained an Act of Parliament enabling him to cut a canal to Manchester, but had been deterred from commencing the work, both by the immense pecuniary outlay which it would have demanded, and the formidable natural difficulties against which at that time there was probably no engineer in the country able to contend. When the idea, however, was now revived, the extraordinary mechanical genius of Brindley had already acquired for him an extensive reputation, and he was applied to by the duke to survey the ground through which the proposed canal would have to be carried, and to make his report upon the practicability of the scheme. New as he was to this species of engineering, Brindley, confident in his own powers, at once undertook to make the desired examination, and, having finished it, expressed his conviction that the ground presented no difficulties which might not be surmounted. On receiving this assurance, the duke at once determined upon commencing the undertaking; and an Act of Parliament having been obtained in 1758, the powers of which were considerably extended by succeeding Acts, the formation of the canal was begun that year.

From the first the duke resolved that, without regard to expense, every part of the work should be executed in the most perfect manner. One of the chief difficulties to be surmounted was that of procuring a sufficient supply of water; and, therefore, that there might be as little of it as possible wasted, it was determined that the canal should be of uniform level throughout, and of course without locks. It had consequently to be carried in various parts of its course both under hills and over wide and deep valleys. The point, indeed, from which it took its commencement was the heart of the coal mountain at Worsley. Here a large basin was formed, in the first place, from which a tunnel of three-quarters of a mile in length had to be cut through the hill. We may just mention, in passing, that the subterraneous course of the water beyond this basin has since been extended in various directions for about thirty miles. After emerging from under ground, the line of the canal was carried forward, as we have stated, by the intrepid engineer, on the same undeviating level, every obstacle that presented itself being triumphed over by his admirable ingenuity, which the difficulties seemed only to render more fertile in happy inventions. Nor did his comprehensive mind ever neglect even the most subordinate departments of the enterprise. The operations of the workmen were everywhere facilitated by new machines of his contrivance; and whatever could contribute to the economy with which the work was carried on, was attended to only less anxiously than what was deemed essential to its completeness. Thus, for example, the materials excavated from one place were employed to form the necessary embankments at another, to which they were conveyed in boats, having bottoms which opened, and at once deposited the load in the place where it was wanted. No part of his task, indeed, seemed to meet this great engineer unprepared. He made no blunders, and never had either to undo anything or to wish it undone; on the contrary, when any new difficulty occurred, it appeared almost as if he had been all along providing for it—as if his other operations had been directed from the first by his anticipation of the one now about to be undertaken.

In order to bring the canal to Manchester, it was necessary to carry it across the Irwell. That river is, and was then, navigable for a considerable way above the place at which the canal comes up to it; and this circumstance interposed an additional difficulty, as, of course, in establishing the one navigation, it was indispensable that the other should not be destroyed or interfered with. But nothing could dismay the daring genius of Brindley. Thinking it, however, due to his noble employer to give him the most satisfying evidence in his power of the practicability of his design, he requested that another engineer might be called in to give his opinion before its execution should be determined on. This person Brindley carried to the spot where he proposed to rear his aqueduct, and endeavoured to explain to him how he meant to carry on the work. But the man only shook his head, and remarked that 'he had often heard of castles in the air, but never before was shown where any of them were to be erected.' The duke, nevertheless, retained his confidence in his own engineer, and it was resolved that the work should proceed. The erection of the aqueduct, accordingly, was begun in September 1760, and on the 17th of July following, the first boat passed over it, the whole structure forming a bridge of above two hundred yards in length, supported upon three arches, of which the centre one rose nearly forty feet above the surface of the river; on which might be frequently beheld a vessel passing along, while another, with all its masts and sails standing, was holding its undisturbed way directly under its keel.

In 1762, an Act of Parliament was, after much opposition, obtained by the duke, for carrying a branch of his canal to communicate with Liverpool, and so uniting that town, by this method of communication, to Manchester. This portion of the canal, which is more than twenty-nine miles in length, is, like the former, without locks, and is carried by an aqueduct over the Mersey, the arch of which, however, is less lofty than that of the one over the Irwell, as the river is not navigable at the place where it crosses. It passes also over several valleys of considerable width and depth. Before this, the usual price of the carriage of goods between Liverpool and Manchester had been twelve shillings per ton by water, and forty shillings by land; they were now conveyed by the canal, at a charge of six shillings per ton, and with all the regularity of land carriage.

In contemplating this great work, we ought not to overlook the admirable manner in which the enterprising nobleman, at whose expense it was undertaken, performed his part in carrying it on. It was his determination, as we have already stated, from the first, to spare no expense on its completion. Accordingly, he devoted to it during the time of its progress nearly the whole of his revenues, denying himself, all the while, even the ordinary accommodations of his rank, and living on an income of four hundred a year. He had even great commercial difficulties to contend with in the prosecution of his schemes, being at one time unable to raise £500 on his bond on the Royal Exchange; and it was a chief business of his agent, Mr. Gilbert, to ride up and down the country to raise money on his grace's promissory notes. It is true that he was afterwards amply repaid for this outlay and temporary sacrifice ; but the compensation that eventually accrued to him he never might have lived to enjoy ; and at all events, he acted as none but extraordinary men do, in thus voluntarily relinquishing the present for the future, and preferring to any dissipation of his wealth on passing and merely personal objects, the creation of this magnificent monument of lasting public usefulness.[1] Nor was it only in the liberality of his expenditure that the duke approved himself a patron worthy of Brindley. He supported his engineer throughout the undertaking with unflinching spirit, in the face of no little outcry and ridicule, to which the imagined extravagance or impracticability of many of his plans exposed him—and that even from those who were generally accounted the most scientific judges of such matters. The success with which these plans were carried into execution is, probably in no slight degree, to be attributed to the perfect confidence with which their author was thus enabled to proceed.

We have entered at the greater length into the history of this undertaking, both because it was the first of a succession of works of the same description, in which the great engineer of whom we are speaking displayed the unrivalled hardihood,

---

[1] Francis, Duke of Bridgewater, died in 1803, at the age of 67, when the ducal title became extinct, and the earldom passed to his cousin, General Egerton. The income arising from his canal property alone was understood to be, at the time of his death, between £50,000 and £80,000 per annum—a large revenue, but not amounting, although we add to it the rents of his other estates, to anything like that assigned to this nobleman, by the writer of his life, in the *Biographie Universelle*, who informs us that the income-tax which he paid every year amounted alone to £110,000 sterling. 'La somme qu'il payait, chaque année, pour sa portion dans la taxe du revenue (*income-tax*) s'élevait seule à 110,000 livres st.' The fact is, that in the returns which he made under the Act imposing the tax in question, the duke estimated his income at that amount. He left at his death, besides his large property in land, about £600,000 in the funds.

originality, and fertility of his genius, and because from it is also to be dated the commencement of that extended canal navigation which now forms so important a part of our means of internal communication in this country. While the Bridgewater canal was yet in progress, Mr. Brindley was engaged by Lord Gower,[1] and the other principal landed proprietors in Staffordshire, to survey a line for another canal, which it was proposed should pass through that county, and, by uniting the Trent and the Mersey, open for it a communication, by water, with both the east and west coast. Having reported favourably of the practicability of this design, and an Act of Parliament having been obtained in 1765 for carrying it into effect, he was appointed to conduct the work. The scheme was one which had been often thought of; but the supposed impossibility of carrying the canal across the tract of elevated country which stretches along the central region of England had hitherto prevented any attempt to execute it. This was, however, precisely such an obstacle as Brindley delighted to cope with; and he at once overcame it, by carrying a tunnel through Harecastle Hill, of two thousand eight hundred and eighty yards in length, at a depth, in some places, of more than two hundred feet below the surface of the earth. This was only one of five tunnels excavated in different parts of the canal, which extends to the length of ninety-three miles, having seventy-six locks, and passing in its course over many aqueducts. Brindley, however, did not live to execute the whole of this great work, which was finished by his brother-in-law, Mr. Henshall, in 1777, about eleven years after its commencement.

[1] Lord Gower married a sister of the Duke of Bridgewater; and his grace left his canal property in Lancashire to his nephew, the Marquess of Stafford.

During the time that these operations, so new in this country, were in progress, the curious crowded to witness them from all quarters, and the grandeur of many of Brindley's plans seems to have made a deep impression upon even his unscientific visitors. A letter which appeared in the newspapers while he was engaged with the Trent and Mersey Canal, gives us a lively picture of the astonishment with which the multitude viewed what he was about. The writer, it will be observed, alludes particularly to the Harecastle tunnel, the chief difficulty in excavating which arose from the nature of the soil it had to be cut through :—' Gentlemen come to view our eighth wonder of the world, the subterranean navigation which is cutting by the great Mr. Brindley, who handles rocks as easily as you would plum-pies, and makes the four elements subservient to his will. He is as plain a looking man as one of the boors of the Peak, or one of his own carters ; but when he speaks all ears listen, and every mind is filled with wonder at the things he pronounces to be practicable. He has cut a mile through bogs, which he binds up, embanking them with stones, which he gets out of other parts of the navigation, besides about a quarter of a mile into the hill Yelden, on the side of which he has a pump, which is worked by water, and a stove, the fire of which sucks through a pipe the damps that would annoy the men who are cutting towards the centre of the hill. The clay he cuts out serves for brick to arch the subterraneous part, which we heartily wish to see finished to Wilden Ferry, when we shall be able to send coals and pots to London, and to different parts of the globe.'

It would occupy too much of our space to detail, however rapidly, the history of the other undertakings of this description to which the remainder of Mr. Brindley's life was devoted.

The success with which the Duke of Bridgewater's enterprising plans for the improvement of his property were rewarded, speedily prompted numerous other speculations of a similar description; and many canals were formed in different parts of the kingdom, in the execution or planning of almost all of which Brindley's services were employed. He himself had become quite an enthusiast in his new profession, as a little anecdote that has been often told of him may serve to show. Having been called on one occasion to give his evidence touching some professional point before a Committee of the House of Commons, he expressed himself, in the course of his examination, with so much contempt of rivers as means of internal navigation, that an honourable member was tempted to ask him for what purpose he conceived rivers to have been created, when Brindley, after hesitating a moment, replied, 'To feed canals.' His success as a builder of aqueducts would appear to have inspired him with almost as fervid a zeal in favour of bridges as of canals, if it be true, as has been asserted, that one of his favourite schemes contemplated the joining of Great Britain to Ireland by a bridge of boats extending from Portpatrick to Donaghadee. This report, however, is alleged to be without foundation by the late Earl of Bridgewater, in a curious work which he published some years ago at Paris, relative to his predecessor's celebrated canal.

Brindley's multiplied labours and intense application rapidly wasted his strength and shortened his life. He died at Turnhurst, in Staffordshire, on the 27th of September 1772, in the fifty-sixth year of his age, having suffered for some years under a hectic fever which he had never been able to get rid of. In his case, as in that of other active spirits, the soul seems to have

'O'er-inform'd its tenement of clay,'

although the actual bodily fatigue to which his many engagements subjected him must doubtless have contributed to wear him out.

No man ever lived more for his pursuit, or less for himself, than Brindley. He had no sources of enjoyment, or even of thought, except in his profession. It is related that, having once, when in London, been prevailed upon to go to the theatre, the unusual excitement so confused and agitated him as actually to unfit him for business for several days, on which account he never could be induced to repeat his visit. His total want of education, and ignorance of literature, left his genius without any other field in which to exercise itself and spend its strength than that which the pursuit of his profession afforded it: its power, even here, would not probably have been impaired, if it could have better sought relaxation in variety; on the contrary, its spring would most likely have been all the stronger for being occasionally unbent. We have already mentioned that he was all but entirely ignorant of reading and writing. He knew something of figures, but did not avail himself much of their assistance in performing the calculations which were frequently necessary in the prosecution of his mechanical designs. On these occasions his habit was to work the question, by a method of his own chiefly in his head, only setting down the results at particular stages of the operation; yet his conclusions were generally correct. His vigour of conception, in regard to machinery was so great, that, however complicated might be the machine he had to execute, he never, except sometimes to satisfy his employers, made any drawing or model of it, but having once fixed its different parts in his mind, would construct it without any difficulty merely from the idea of which he had thus possessed himself. When much perplexed with any problem he had to solve, his practice was to take to bed in order to study it, and he would sometimes remain, we are told, for two or three days thus fixed to his pillow in meditation.

We shall the more clearly appreciate the impulse given to inland navigation in this country by the achievements of Brindley, and the extent of the new accommodation which our commerce has hence obtained within the last sixty or seventy years, if we cast our eye for a moment over the map of Great Britain, and note a few of the principal canals by which the island is now intersected in all directions. First, there is the Trent and Mersey Canal, which we have already mentioned, and which was denominated by Brindley the Grand Trunk Navigation, as, in fact, uniting one side of the kingdom to the other, and therefore specially adapted to serve, as it has since actually done, by way of stem from which other similar lines might proceed as branches to different points. By this canal, a complete water communication was established, though by a somewhat circuitous sweep, between the great ports of Liverpool on the west coast and Hull on the east. A branch from it, the Staffordshire and Worcestershire Canal, was afterwards carried to the river Severn; and thus a union was effected between the port of Bristol and the two already mentioned. This branch, being about forty-six miles long, was also executed by Brindley, and was completed in 1772. Similar communications were subsequently formed from other points on the south coast to the central counties. But the most important line of English canals is that which extends from the centre of the kingdom to the metropolis, and, by falling into the Grand Trunk Navigation, forms in fact a continued communication by water all the way from London to Liverpool. Of this line, the principal part is formed by what is called the Grand Junction Canal, which, commencing at Brentford, stretches north-west till it falls into a branch of the Oxford Canal at Braunston, in Northamptonshire, passing at one place (Blisworth) through a tunnel three thousand and eighty yards in length, eighteen feet high, and sixteen and a half wide. The Regent and Paddington Canals have since formed communications between the Grand Junction Canal and the eastern, western, and northern parts of the metropolis. The whole length of the direct waterway thus established between Liverpool and London is about two hundred and sixty-four miles; but if the different canals which contribute to form the line be all of them measured in their entire length, the aggregate amount of the inland navigation, in this connection alone, will be found to extend to above one thousand four hundred miles.

The oldest canal in the northern part of the kingdom is that between the Forth and Clyde, which was executed by the celebrated Smeaton, although its plan was revised by Brindley. It commences at Grangemouth, on the Carron, at a short distance from where that river falls into the Forth, and originally terminated at Port Dundas, in the neighbourhood of Glasgow. A portion of this canal, owing to the great descent of the ground over which it passes towards the west, has no fewer than twenty locks in the first ten miles and a half. It was afterwards carried farther west to Dalmuir, on the Clyde, and is now connected with the Glasgow and Saltcoats Canal, whose course is across the counties of

Renfrew and Ayr to the river Garnock, which flows into the Atlantic opposite to the Isle of Arran. More recently, a branch was extended from its north-eastern extremity, along the south bank of the Forth, as far as Edinburgh, so that the whole now forms an uninterrupted line of canal navigation from the east to the west coast of Scotland. The famous Caledonian Canal, in the north of Scotland, also unites the two opposite seas, and indeed runs pretty nearly parallel to a part of the line that has just been described. It was commenced in 1802, under the management of Mr. Telford, who conducted it throughout, and was first opened on the 23d of October 1822. The distance between the German and the Atlantic Oceans, measured in the direction of this canal, is two hundred and fifty miles; but of this nearly two hundred and thirty miles, consisting of friths and lakes, were already navigable. The canal itself, therefore, which has cost about a million of pounds sterling, is only, properly speaking, about twenty miles in length; and, had not steam navigation been fortunately discovered while the work was going on, there seems every reason to believe that the cut would have been nearly useless.

The entire length of the canal navigation already formed in Great Britain and Ireland exceeds four thousand seven hundred miles. The whole of this is the creation of less than a century, during which period, therefore, considerably above forty miles of canal may be said to have been produced every year—a truly extraordinary evidence of the spirit and resources of a country which has been able to continue so large an expenditure for so long a time on a single object, and which has in a single year, during that period, spent almost as much money upon war as all those canals together have cost for three-quarters of a century. If Brindley had never lived, we should undoubtedly ere now have been in possession of much of this accommodation; for the time was ripe for its introduction, and an increasing commerce, everywhere seeking vent, could not have failed, ere long, to have struck out for itself, to a certain extent, these new facilities. But had it not been for the example set by his adventurous genius, the progress of artificial navigation among us would probably have been timid and slow compared to what it has been. For a long time, in all likelihood, our only canals would have been a few small ones, cut in the more level parts of the country, like that substituted in 1755 for the Sankey Brook, the benefit of each of which would have been extremely insignificant and confined to a very narrow neighbourhood. He did, in the very infancy of the art, what has not yet been undone, struggling, indeed, with such difficulties, and triumphing over them, as could be scarcely exceeded by any his successors might have to encounter. By the boldness and success with which, in particular, he carried the Grand Trunk Navigation across the elevated ground of the Midland Counties, he demonstrated that there was hardly any part of the island where a canal might not be formed; and, accordingly, this very central ridge, which used to be deemed so insurmountable an obstacle to the junction of our opposite coasts, is now intersected by more than twenty canals beside the one which he first drove through the barrier. It is in the conception and accomplishment of such grand and fortunate deviations from ordinary practice that we discern the power, and confess the value, of original genius. The case of Brindley affords us a wonderful example of what the force of natural talent will sometimes do in attaining an acquaintance with particular departments of science, in the face of almost every conceivable disadvantage, where not only all education is wanting, but even all access to books.

WILLIAM COBBETT.

WILLIAM COBBETT was a native of Farnham, in Surrey. He was born about 1762, the third son of a small farmer. After he had risen to eminence and distinction, it was his delight and his pride to refer to the honourable, if humble circumstances of his early life—to a father whom, he says, 'I ardently loved, and to whose every word I listened with admiration,' and to a 'gentle, and tender-hearted, and affectionate mother.' In one of his *Rural Rides*, in which he was accompanied by one of his sons, then a mere boy, he says: 'In coming from Moor Park to Farnham town, I stopped opposite the door of a little old house, where there appeared to be many children. "There, Dick," said I, "when I was just such a little creature as that, whom you see in the doorway, I lived in this very house with my grandmother Cobbett."' He was a bold, adventurous, hardy little chap, fond of all manner of rural English sports, and the very 'father to the man' he afterwards became. Cobbett, whatever were his faults, had a genial temperament and great warmth of feeling. In one of his *Rural Rides*, in which he was accompanied by an elder son, he writes:—

'We went a little out of the way to go to a place called the Bourne, which lies in the heath at about a mile from Farnham. We went to Bourne in order that I might show my son the spot where I received the rudiments of my education. There is a little hop garden in which I used to work when from eight to ten years old, from which I have scores of times run to follow the hounds, leaving the hoe to do the best that it could to destroy the weeds. But the most interesting thing was a sand-hill which goes from a part of the heath down to the rivulet. As a due mixture of pleasure with toil, I, with two brothers, used occasionally to disport ourselves, as the lawyers call it, at this sand-hill. Our diversion was this. We used to go to the top of the hill, which was steeper than the roof of a house; one used to draw his arms out of the sleeves of his smock-frock, and lay himself down with his arms by his sides; and then the others, one at head and the other at feet, sent him rolling down the hill like a barrel or a log of wood. By the time he got to the bottom, his hair, eyes, ears, nose, and mouth were all full of this loose sand; then the others took their turn, and, at every roll, there was a monstrous spell of laughter. I had often told my sons of this, while they were very little, and I now took one of them to see the spot. But that was not all. This was the spot where I was receiving my education; and this was the sort of education; and I am perfectly satisfied that, if I had not received such an education, or something very much like it,—that, if I had been brought up a milksop, with a nursery-maid everlastingly at my heels,—I should have been at this day as great a fool, as inefficient a mortal, as any of those frivolous idiots that are turned out from Winchester and Westminster School, or from any of those dens of dunces called colleges and universities. It is impossible to say how much I owe to that sand-hill; and I went to return it my thanks for the ability which it probably gave me to be one of the greatest terrors to one of the greatest and most powerful bodies of knaves and fools that ever were permitted to afflict this or any other country.'

Breakfasting at a little village in Sussex, he looks with fond complacency upon the landlady's son: 'A very pretty village, and a very nice breakfast, in a very neat parlour of a very decent public-house. The landlady sent her son to get me some cream; and he was just such a chap as I was at his age, and dressed just in the same sort of way, his main garment being a blue smock-frock, faded from wear, and mended with pieces of new stuff, and, of course, not faded. The sight of this smock-frock brought to my recollection many things very dear to me.' This is as fine as Burns gazing upon the cottage smoke in his morning walk to Blackford Hill with Dugald Stewart. One anecdote of his boyhood, related by himself, is so amusingly characteristic of the future man, that we have never forgotten it. He was not permitted to follow the hounds upon some occasion, and, in revenge, procured a salt herring, which he furtively drew over the ground where they were to throw off, thus to cast them off the scent. The trick took to admiration, and the boy as much exulted in his success as did the man in the discomfiture of his enemies, Ellenborough and Vickary Gibbs.

In the introduction to one of his most delightful books,—next, indeed, to the *Rural Rides*,—namely, his *Year's Residence in America*, he says:—

'Early habits and affections seldom quit us while we have vigour of mind left. I was brought up under a father whose talk was chiefly about his garden and his fields, with regard to which he was famed for his skill and his exemplary neatness. From my very infancy, from the age of six years, when I climbed up the side of a steep sand-rock, and there scooped me out a plot four feet square to make me a garden, and the soil for which I carried up in the bosom of my little blue smock-frock, or hunting-shirt, I have never lost one particle of my passion for these healthy, and rational, and heart-cheering pursuits, in which every day presents something new, in which the spirits are never suffered to flag, and in which industry, skill, and care are sure to meet with their due reward. I have never, for any eight months together, during my whole life, been without a garden.'

In the same volume in his American journal, this passage occurs :—

'When I returned to England in 1800, after an absence, from the country parts of it, of sixteen years, the trees, the hedges, even the parks and woods, seemed so small! It made me laugh to hear little gutters, that I could jump over, called rivers! The Thames was but a "creek"! But when, in about a month after my arrival in London, I went to Farnham, the place of my birth, what was my surprise! Everything was become so pitifully small! I had to cross, in my post-chaise, the long and dreary heath of Bagshot; then, at the end of it, to mount a hill called Hungry Hill; and from that hill I knew that I should look down into the beautiful and fertile vale of Farnham. My heart fluttered with impatience, mixed with a sort of fear, to see all the scenes of my childhood; for I had learnt before, the death of my father and mother. There is a hill, not far from the town, called Crooksbury Hill, which rises up out of a flat, in the form of a cone, and is planted with Scotch fir-trees. Here I used to take the eggs and young ones of crows and magpies. This hill was a famous object in the neighbourhood. It served as the superlative degree of height. "As high as Crooksbury Hill" meant with us the utmost degree of height. Therefore, the first object that my eyes sought was this hill. *I could not believe my eyes!* Literally speaking, I for a moment thought the famous hill removed, and a little heap put in its stead; for I had seen, in New Brunswick, a single rock, or hill of solid rock, ten times as big, and four or five times as high! The postboy, going down hill, and not a bad road, whisked me, in a few minutes, to the Bush Inn, from the garden of which I could see the prodigious sand-hill where I had begun my gardening works. What a nothing! But now came rushing into my mind, all at once, my pretty little garden, my little blue smock-frock, my little nailed shoes, my pretty pigeons that I used to feed out of my hands, the last kind words and tears of my gentle, and tender-hearted, and affectionate mother! I hastened back into the room. If I had looked a moment longer, I should have dropped. When I came to reflect, what a change! I looked down at my dress. What a change! What scenes I had gone through! How altered my state! I had dined the day before at the Secretary of State's, in company with Mr. Pitt, and had been waited upon by men in gaudy liveries! I had had nobody to assist me in the world. No teachers of any sort. Nobody to shelter me from the consequences of bad, and no one to counsel me to good behaviour. I felt proud. The distinctions of rank, birth, and wealth, all became nothing in my eyes, and from that moment (less than a month after my arrival in England) I resolved never to bend before them.'

Cobbett, in his native place, and following the employments of his ancestors, must inevitably have been a 'village Hampden.' On looking at a little smock-frocked boy, in nailed shoes and clean coarse shirt, such as he had been, he very naturally remarks: 'If accident had not taken me from a similar scene, how many villains and fools, who have been well teased and tormented, would have slept in peace by night, and fearlessly swaggered about by day!' Cobbett received so little school learning that, in his case, it may be almost truly said 'reading and writing came by nature.' From eight years of age he was engaged in such rural occupations as picking hops and hautboys, weeding in gardens, and driving away the birds, and following the hounds, or getting upon horseback as often as he could, or digging after rabbits' nests, rolling down the sand-hills, and whipping the little *efts* that crept about in the heath. And this is the education which, upon reflection, he preferred. None of his own young children were ever sent from home to school. Reading and writing came to them from imitation. Throughout all Cobbett's writings (crotchets notwithstanding), excellent hints are scattered upon this important subject, but especially in his *Advice to Young Men*. His controversy with the educators as a sect, was merely one of sound. No man could prize the advantages of education so highly as one who owed all he knew to himself, and who had pursued knowledge unremittingly, and under considerable difficulties. His first start from home he has described himself in this memorable passage:—

'At eleven years of age, my employment was clipping off box-edgings and weeding beds of flowers in the garden of the Bishop of Winchester, at the Castle of Farnham, my native town. I had always been fond of beautiful gardens, and a gardener, who had just come from the King's Gardens at Kew, gave such a description of them as made me instantly resolve to work in these gardens. The next morning, without saying a word to any one, off I set, with no clothes except those upon my back, and with thirteen halfpence in my pocket. I found that I must go to Richmond, and I accordingly went on from place to place, inquiring my way thither. A long day (it was in June) brought me to Richmond in the afternoon. Two pennyworth of bread and cheese and a pennyworth of small beer which I had on the road, and one halfpenny which I had lost somehow or other, left threepence in my pocket. With this for my whole fortune, I was trudging through Richmond in my blue smock-frock and my red garters tied under my knees, when, staring about me, my eye fell upon a little book in a bookseller's window, on the outside of which was written, *Tale of a Tub*, price 3d. The title was so odd that my curiosity was excited. I had the threepence, but then I could have no supper. In I went and got the little book, which I was so impatient to read that I got over into a field at the upper corner of the Kew Garden, where there stood a hay-stack. On the shady side of this I sat down to read. The book was so different from anything that I had read before; it was something so new to my mind that, though I could not at all understand some of it, it delighted me beyond description, and it produced what I have always considered a sort of birth of intellect. I read on till it was dark, without any thought about supper or bed. When

I could see no longer, I put my little book in my pocket and tumbled down by the side of the stack, where I slept till the birds in Kew Gardens awaked me in the morning, when off I started to Kew, reading my little book. The singularity of my dress, the simplicity of my manner, my confident and lively air, and, doubtless, his own compassion besides, induced the gardener, who was a Scotsman, to give me victuals, find me lodging, and set me to work. And it was during the period that I was at Kew that the present king and two of his brothers laughed at the oddness of my dress while I was sweeping the grass plat round the foot of the pagoda. The gardener, seeing me fond of books, lent me some gardening books to read; but these I could not relish after my *Tale of a Tub*, which I carried about with me wherever I went; and when I, at about twenty years old, lost it in a box that fell overboard in the Bay of Fundy, in North America, the loss gave me greater pain than I have ever felt at losing thousands of pounds. This circumstance, trifling as it was, and childish as it may seem to relate it, has always endeared the recollection of Kew to me.'

At sixteen he attempted to make off to sea; at seventeen he went to London, where he supported himself for some time as a copying clerk; at twenty-two he enlisted as a private soldier, and rose to the rank of sergeant-major. His regiment was the 53d, then commanded by one of the king's sons, the Duke of Kent, and he went with it to British America. Thus, from a very tender age he was left entirely to his own guidance and mastership, and thus was nourished the self-depending, determined character which nerved him for his lifelong struggle. The little illustrative snatches of personal history, especially of his young days, which he has incidentally given, are the most attractive part of his writings, and these, fortunately, mingle the most largely in the more popular and enduring part of them, namely—the *Rural Rides*, the *Year's Residence in America*, and the *Advice to Young Men*. In the latter work he says, in treating of education, and, in particular, of learning grammar :—

'The study need subtract from the hours of no business, nor, indeed, from the hours of necessary exercise; the hours usually spent on the tea and coffee slops, and in the mere gossip which accompany them—those wasted hours of only one year employed in the study of English grammar would make you a correct speaker and writer for the rest of your life. You want no school, no room to study in, no expenses, and no troublesome circumstances of any sort. I learned grammar when I was a private soldier on the pay of sixpence a day. The edge of my berth or that of the guard-bed was my seat to study in, my knapsack was my bookcase, a bit of board lying on my lap was my writing-table, and the task did not demand anything like a year of my life. I had no money to purchase candle or oil; in winter time it was rarely that I could get any evening light but that of the fire, and only my turn even of that. And if I under such circumstances, and without parent or friend to advise or encourage me, accomplished this undertaking, what excuse can there be for any youth, however poor, however pressed with business, or however circumstanced as to room or other conveniences? To buy a pen or a sheet of paper, I was compelled to forego some portion of food, though in a state of half starvation; I had no moment of time that I could call my own, and I had to read and to write amidst the talking, laughing, singing, whistling, and brawling of at least half a score of the most thoughtless of men, and that, too, in the hours of their freedom from all control. Think not lightly of the farthing that I had to give now and then for ink, pen, or paper. That farthing was, alas! a great sum to me. I was as tall as I am now; I had great health and great exercise. The whole of the money not expended for us at market was twopence a week for each man. I remember, and well I may, that upon one occasion I, after all absolutely necessary expenses, had, on a Friday, made shift to have a halfpenny in reserve, which I had destined for the purchase of a red herring in the morning; but when I pulled off my clothes at night, so hungry then as to be hardly able to endure life, I found that I had lost my halfpenny. I buried my head under the miserable sheet and rug, and cried like a child. And again I say, if I, under circumstances like these, could encounter and overcome this task, is there, can there be, in the whole world a youth to find an excuse for the non-performance? What youth who shall read this will not be ashamed to say that he is not able to find time and opportunity for this most essential of all the branches of book-learning!'

His natural disposition, prompt and active, made him fall easily into the better parts of military habits. The original maxim of the man who for forty years daily did so much, and who, having put his hand to the plough, never once looked back, was *Toujours prêt*, 'always ready;' and it ought to be the family motto of the Cobbetts. He says of himself:—

'For my part, I can truly say that I owe more of my great labours to my strict adherence to the precepts that I have here given you than to all the natural abilities with which I have been endowed; for these, whatever may have been their amount, would have been of comparatively little use, even aided by great sobriety and abstinence, if I had not in early life contracted the blessed habit of husbanding well my time.

To this, more than to any other thing, I owed my very extraordinary promotion in the army. I was "always ready;" if I had to mount guard at ten, I was ready at nine; never did any man, or anything, wait one moment for me. Being, at an age under twenty years, raised from corporal to sergeant-major at once, over the heads of thirty sergeants, I naturally should have been an object of envy and hatred; but this habit of early rising and of rigid adherence to the precepts which I have given you really subdued these passions, because every one felt that what I did he had never done, and never could do. Before my promotion, a clerk was wanted to make out the morning report of the regiment. I rendered the clerk unnecessary, and long before any other man was dressed for the parade, my work for the morning was all done, and I myself was on the parade, walking, in fine weather, for an hour perhaps. My custom was this: to get up in summer at daylight, and in winter at four o'clock—shave, dress, even to the putting of my sword-belt over my shoulder, and having my sword lying on the table before me ready to hang by my side. Then I ate a bit of cheese, or pork and bread. Then I prepared my report, which was filled up as fast as the companies brought me in the materials. After this I had an hour or two to read before the time came for any duty out of doors, unless when the regiment, or part of it, went out to exercise in the morning. When this was the case, and the matter was left to me, I always had it on the ground in such time as that the bayonets glistened in the rising sun—a sight which gave me delight, of which I often think, but which I should in vain endeavour to describe. If the officers were to go out, eight or ten o'clock was the hour, sweating the men in the heat of the day, breaking in upon the time for cooking their dinner, putting all things out of order and all men out of humour. When I was commander, the men had a long day of leisure before them; they could ramble into the town or into the woods, go to get raspberries, to catch birds, to catch fish, or to pursue any other recreation, and such of them as chose and were qualified, to work at their trades.'

Much of the spare time of Cobbett was, in his younger years, devoted to a very miscellaneous kind of reading. He ran through all the books of a country circulating library, trash and all; and, contemptibly as he often affects to speak of literary pursuits, the fruits of these early studies are often revealed in the lively style and the fertility and happiness of allusion which distinguish all his writings. No one has abused Shakespeare so absurdly and truculently—for this was one of Cobbett's many crotchets; but, then, few have quoted the bard of many-coloured life so aptly and frequently. Shakespeare and the principal English poets were clearly at his finger ends, while, from wayard caprice, he affected ignorance, with contempt of them. Of the arts he knew nothing, not even the mechanic arts; and his tours in Scotland and Ireland show how little he possessed of what is called general information—the kind of knowledge which comes almost of itself, and which he despised much more than was needful. Yet, his acquaintance with English classical literature, and even with contemporary authors, must have been extensive, and gradually accumulating, in the gardens of Kew in London, and in New Brunswick, and to the last hour of his life. The *Tale of a Tub* had introduced the boy to the writings of Swift; and we have been informed by an officer who joined the 53d Regiment shortly after Cobbett left it, that he had written out in some of the regimental books, *Directions for a Sergeant*

Major, or an orderly, in the manner of Swift's *Advice to Servants*, which were full of admirable humour and grave irony. The officers of the 53d and the corps were, as we have reason to know, exceedingly proud of their clever sergeant-major after he became famous; and so, indeed, was the whole army, from the period he became a party writer in Philadelphia. He was particularly distinguished by his Royal Highness the Duke of Kent.

In the *Advice to Young Men*, which may be called his confessions, Cobbett has related his own love story; and a delightful one it is, possessing at once the tenderness and simplicity of nature, and no little of the charm of romance. The scene of it was New Brunswick. But there is a collateral flirtation also, involving what Cobbett terms the only serious sin he ever committed against the female sex, and which he relates in warning to young men. We shall take it first, and that, too, in the language of his own narrative.

'The province of New Brunswick, in North America, in which I passed my years from the age of eighteen to that of twenty-six, consists, in general, of heaps of rocks, in the interstices of which grow the pine, the spruce, and various sorts of fir trees, or, where the woods have been burnt down, the bushes of the raspberry or those of the huckle-berry. The province is cut asunder lengthwise by a great river, called the St. John, about two hundred miles in length and at half-way from the mouth, fully a mile wide. Into this main river run innumerable smaller rivers, there called creeks. On the sides of these creeks the land is, in places, clear of rocks; it is in these places generally good and productive: the trees that grow here are the birch, the maple, and others of the deciduous class; natural meadows here and there present themselves; and some of these spots far surpass in rural beauty any other that my eyes ever beheld, the creeks abounding towards their sources in waterfalls of endless variety, as well in form as in magnitude, and always teeming with fish; while waterfowl enliven their surface, and wild pigeons, of the gayest plumage, flutter, in thousands upon thousands, amongst the branches of the beautiful trees, which, sometimes for miles together, form an arch over the creeks.

'I, in one of my rambles in the woods, in which I took great delight, came to a spot at a very short distance from the source of one of these creeks. Here was everything to delight the eye, and especially of one like me, who seems to have been born to love rural life, and trees and plants of all sorts. Here was about two hundred acres of natural meadow, interspersed with patches of maple trees, in various forms and of various extent; the creek came down in cascades, for any one of which many a nobleman in England would, if he could transfer it, give a good slice of his fertile estate; and, in the creek, at the foot of the cascades, there were, in the season, salmon the finest in the world, and so abundant and so easily taken as to be used for manuring the land.

'If Nature, in her very best humour, had made a spot for the express purpose of captivating me, she could not have exceeded the efforts which she had here made. But I found something here besides these rude works of nature; I found something in the fashioning of which man had had something to do. I found a large and well-built log dwelling-house, standing (in the month of September) on the edge of a very good field of Indian corn, by the side of which there was a piece of buckwheat just then mowed. I found a homestead, and some very pretty cows. I found all the things by which an easy and happy farmer is surrounded; and I found still something besides all these—something that was destined to give me a great deal of pleasure and also a great deal of pain, both in their extreme degree, and both of which, in spite of the lapse of forty years, now make an attempt to rush back into my heart.

'Partly from misinformation, and partly from miscalculation, I had lost my way; and, quite alone, but armed with my sword and a brace of pistols, to defend myself against the bears, I arrived at the log-house in the middle of a moonlight night, the hoar-frost covering the trees and the grass. A stout and clamorous dog, kept off by the gleaming of my sword, waked the master of the house, who got up, received me with great hospitality, got me something to eat, and put me into a feather bed, a thing that I had been a stranger to for some years. I, being very tired, had tried to pass the night in the woods, between the trunks of two large trees which had fallen side by side, and within a yard of each other. I had made a nest for myself of dry fern, and had made a covering by laying boughs of spruce across the trunks of the trees. But, unable to sleep on account of the cold, becoming sick from the great quantity of water that I had drunk during the heat of the day, and being, moreover, alarmed at the noise of the bears, and lest one of them should find me in a defenceless state, I had roused myself up and had crept along as well as I could; so that no hero of eastern romance ever experienced a more enchanting change.

'I had got into the house of one of those Yankee loyalists, who, at the close of the revolutionary war (which, until it had succeeded, was called a rebellion), had accepted of grants of land in the king's province of New Brunswick, and who, to the great honour of England, had been furnished with all the means of making new and comfortable settlements. I was suffered to sleep till breakfast time, when I found a table, the like of which I have since seen so many times in the United States, loaded with good things. The master and the mistress of the house, aged about fifty, were like what an English farmer and his wife were half a century ago. There were two sons, tall and stout, who appeared to have come in from work, and the youngest of whom was about my age, then twenty-three. But there was another member of the family, aged nineteen, who (dressed according to the neat and simple fashion of New England, whence she had come with her parents five or six years before) had her long light-brown hair twisted nicely up, and fastened on the top of her head, in which head were a pair of lively blue eyes, associated with features of which that softness and that sweetness so characteristic of American girls were the predominant expressions, the whole being set off by a complexion indicative of glowing health, and forming—figure, movements, and all taken together—an assemblage of beauties far surpassing any that I had ever seen but once in my life. That once was, too, two years agone; and, in such a case and at such an age, two years, two whole years, is a long, long while! It was a space as long as the eleventh part of my then life! Here was the present against the absent; here was the power of the eyes pitted against that of the memory; here were all the senses up in arms to subdue the influence of the thoughts; here was vanity, here was passion, here was the spot of all spots in the world, and here were also the life, and the manners, and the habits, and the pursuits that I delighted in; here was everything that imagination can conceive — united in a conspiracy against the poor little brunette in England! What, then, did I fall in love at once with this bouquet of lilies and roses? Oh! by no means. I was, however, so enchanted with the place, I so much enjoyed its tranquillity, the shade of the maple trees, the business of the farm, the sports of the water and of the woods, that I stayed at it to the last possible minute, promising, at my departure, to come again as often as I possibly could—a promise which I most punctually fulfilled.

'Winter is the great season for jaunting and dancing (called frolicking) in America. In this province, the river and the creeks were the only roads from settlement to settlement. In summer we travelled in canoes; in winter, in sleighs on the ice or snow. During more than two years, I spent all the time I could with my Yankee friends: they were all fond of me: I talked to them about country affairs, my evident delight in which they took as a compliment to themselves: the father and mother treated me as one of their children, the sons as a brother, and the daughter, who was as modest and as full of sensibility as she was beautiful, in a way to which a chap much less sanguine than I was, would have given the tenderest interpretation; which treatment I, especially in the last-mentioned case, most cordially repaid.

'Yet I was not a deceiver; for my affection for her was very great: I spent no really pleasant hours but with her: I was uneasy if she showed the slightest regard for any other young man: I was unhappy if the smallest matter affected her health or spirits: I quitted her in dejection, and returned to her with eager delight: many a time, when I could get leave but for a day, I paddled in a canoe two whole succeeding nights, in order to pass that day with her. If this was not love, it was first cousin to it; for, as to any criminal intention, I no more thought of it, in her case, than if she had been my sister. Many times I put to myself the questions, "What am I at? Is not this wrong? Why do I go?" But still I went.

'The last parting came; and now came my just punishment! The time was known to everybody, and was irrevocably fixed; for I had to move with a regiment, and the embarkation of a regiment is an epoch in a thinly-settled province. To describe this parting would be too painful even at this distant day, and with this frost of age upon my head. The kind and virtuous father came forty miles to see me just as I was going on board in the river. His looks and words I have never forgotten. As the vessel descended, she passed the mouth of that creek which I had so often entered with delight; and though England, and all that England contained, were before me, I lost sight of this creek with an aching heart.

'On what trifles turn the great events in the life of man! If I had received a cool letter from my intended wife; if I had only heard a rumour of anything from which fickleness in her might have been inferred; if I had found in her any, even the smallest, abatement of affection; if she had but let go any one of the hundred strings by which she held my heart: if any one of these, never would the world have heard of me. Young as I was; able as I was as a soldier; proud as I was of the admiration and commendations of which I was the object; fond as I was, too, of the command which, at so early an age, my rare conduct and great natural talents had given me; sanguine as was my mind, and brilliant as were my prospects: yet I had seen so much of the meannesses, the unjust partialities, the insolent pomposity, the disgusting dissipations of that way of life, that I was weary of it: I longed exchanging my fine-laced coat for the Yankee farmer's home-spun, to be where I should never behold the supple crouch of servility, and never hear the hectoring voice of authority again; and, on the lonely banks of this branch-covered creek, which contained (she out of the question) everything congenial to my taste and dear to my heart, I, unapplauded, unfeared, unenvied, and uncalumniated, should have lived and died.'

The fair cause of this 'serious sin,' the little brunette in England, had first been seen some years before in America, and after this charming manner: 'When I first saw my wife, she was thirteen years old, and I was within about a month of twenty-one. She was the daughter of a sergeant of artillery, and I was the sergeant-major of a regiment of foot, both stationed in forts near the city of St. John, in the province of New Brunswick. I sat in the same room with her for about an hour, in company with others, and I made up my mind that she was the very girl for me. That I thought her beautiful, is certain; for that, I had always said, should be an indispensable qualification; but I saw in her what I deemed marks of that sobriety of conduct of which I have said so much, and which has been by far the greatest blessing of my life. It was now dead of winter, and, of course, the snow several feet deep on the ground, and the weather piercing cold. It was my habit, when I had done my morning's writing, to go out at break of day to take a walk on a hill, at the foot of which our barracks lay. In about three mornings after I had first seen her, I had, by an invitation to breakfast with me, got up two young men to join me in my walk; and our road lay by the house of her father and mother. It was hardly light, but she was out on the snow scrubbing out a washing-tub. "That's the girl for me," said I, when we had got out of her hearing. One of these young men came to England soon afterwards; and he, who keeps an inn in Yorkshire, came over to Preston, at the time of the election, to verify whether I were the same man. When he found that I was, he appeared surprised; but what was his surprise when I told him that those tall young men, whom he saw around me, were the sons of that pretty little girl that he and I saw scrubbing out the washing-tub on the snow in New Brunswick at daybreak.

'From the day that I first spoke to her, I never had a thought of her being the wife of any other man, more than I had a thought of her being transformed into a chest of drawers; and I formed my resolution at once, to marry her as soon as we could get permission, and to get out of the army as soon as I could. So that this matter was at once settled as firmly as if written in the book of fate. At the end of about six months, my regiment, and I along with it, were removed to Frederickton, a distance of a hundred miles up the river of St. John; and, which was worse, the artillery were expected to go off to England a year or two before our regiment! The artillery went, and she along with them; and now it was that I acted a part becoming a real and sensible lover. I was aware that, when she got to that gay place, Woolwich, the house of her father and mother, necessarily visited by numerous persons, not the most select, might become unpleasant to her; and I did not like, besides, that she should continue to work hard. I had saved a hundred and fifty guineas, the earnings of my early hours, in writing for the paymaster, the quartermaster, and others, in addition to the savings of my own pay. I sent her all my money before she sailed, and wrote to her to beg of her, if she found her home uncomfortable, to hire a lodging with respectable people, and, at any rate, not to spare the money by any means, but to buy herself good clothes, and to live without hard work until I arrived in England; and I, in order to induce her to lay out the money, told her that I should get plenty more before I came home.

'We were kept abroad two years longer than our time, Mr. Pitt (England not being so tame then as she is now) having knocked up a dust with Spain about Nootka Sound. Oh, how I cursed Nootka Sound, and poor bawling Pitt, too, I am afraid! At the end of four years, however, home I came, landed at Portsmouth, and got my discharge from the army, by the great kindness of poor Lord Edward Fitzgerald, who was then the major of my regiment. I found my little girl a servant of all work (and hard work it was), at five pounds a year, in the house of a Captain Brisac, and, without hardly saying a word about the matter, she put into my hands the whole of my hundred and fifty guineas unbroken!

'Need I tell the reader what my feelings were? Need I tell kind-hearted English parents what effect this anecdote must have produced on the minds of our children?'

After his marriage, Cobbett lived with his wife for some time in France, studying the language; and then they went to Philadelphia, where he began to teach English to Frenchmen, and, as his first work, composed his French and English grammar. He remained between Philadelphia and New York for about eight years, and, during most of this time, had a printing establishment and a book store.

In the *Advice to Young Men*, he pictures his domestic character and habits at this period in the most engaging manner, and, we daresay, not too much *en beau*, for all is so simply right and so perfectly natural. But this, as has been remarked, is the sanctified life of the fireside—' the porcupine with his quills sheathed.' He says:—

'I began my young marriage days in and near Philadelphia. At one of those times to which I have just alluded, in the middle of the burning hot month of July, I was greatly afraid of fatal consequences to my wife for want of sleep, she not having, after the great danger was over, had any sleep for more than forty-eight hours. All great cities in hot countries are, I believe, full of dogs; and they in the very hot weather keep up during the night a horrible barking and fighting and howling. Upon the particular occasion to which I am adverting, they made a noise so terrible and so unremitted, that it was next to impossible that even a person in full health and free from pain should obtain a minute's sleep. I was, about nine in the evening, sitting by the bed. "I do think," said she, "that I could go to sleep now, if it were not for the dogs." Down-stairs I went, and out I sallied in my shirt and trousers, and without shoes and stockings; and going to a heap of stones lying beside the road, set to work upon the dogs, going backward and forward and keeping them at two or three hundred yards' distance from the house. I walked thus the whole night, barefooted, lest the noise of my shoes might possibly reach her ears; and I remember that the bricks of the causeway were, even in the night, so hot as to be disagreeable to my feet. My exertions produced the desired effect: a sleep of several hours was the consequence, and at eight o'clock in the morning off went I to a day's business which was to end at six in the evening.

'Women are all patriots of the soil, and when her neighbours used to ask my wife whether all English husbands were like hers, she boldly answered in the affirmative. I had business to occupy the whole of my time, Sundays and week days, except sleeping hours; but I used to make time to assist her in the taking care of her baby, and in all sorts of things —get up, light her fire, boil her tea-kettle, carry her up warm water in cold weather, take the child while she dressed herself and got the breakfast ready, then breakfast, get her in water and wood for the day, then dress myself neatly and sally forth to my business. The moment that was over, I used to hasten back to her again, and I no more thought of spending a moment away from her, unless business compelled me, than I thought of quitting the country and going to sea. The thunder and lightning are tremendous in America compared with what they are in England. My wife was at one time very much afraid of thunder and lightning, and as is the feeling of all such women, and indeed all men too, she wanted company, and particularly her husband, in those times of danger. I knew well, of course, that my presence would not diminish the danger; but be I at what I might, if within reach of home, I used to quit my business and hasten to her the moment I perceived a thunderstorm approaching. Scores of miles have I, first and last, run on this errand in the streets of Philadelphia. The Frenchmen who were my scholars used to laugh at me exceedingly on this account, and sometimes when I was making an appointment with them they would say, with a smile and a bow, "*Sauve la tonnere toujours, Monsieur Cobbett.*"

'I never dangled about at the heels of my wife; seldom,

very seldom, ever walked out, as it is called, with her; I never "went a-walking" in the whole course of my life, never went to walk without having some object in view other than the walk, and as I never could walk at a slow pace, it would have been hard work for her to keep up with me.'

There is much plain sense and manly tenderness to be found in this volume of confessions. This is for the rapidly-increasing sect of club frequenters :—

'What are we to think of the husband who is in the habit of leaving his own fireside, after the business of the day is over, and seeking promiscuous companions in the ale or the coffee house? I am told that in France it is rare to meet with a husband who does not spend every evening of his life in what is called a *café*—that is to say, a place for no other purpose than that of gossiping, drinking, and gaming. And it is with great sorrow that I acknowledge that many English husbands indulge too much in a similar habit. Drinking clubs, smoking clubs, singing clubs, clubs of oddfellows, whist clubs, sotting clubs—these are inexcusable, they are censurable, they are at once foolish and wicked, even in single men; what must they be, then, in husbands? And how are they to answer, not only to their wives, but to their children, for this profligate abandonment of their homes, this breach of their solemn vow made to the former, this evil example to the latter?

'Innumerable are the miseries that spring from this cause. The expense is, in the first place, very considerable. I much question whether, amongst tradesmen, a shilling a night pays the average score, and that, too, for that which is really worth nothing at all, and cannot, even by possibility, be attended with any one single advantage, however small. Fifteen pounds a year thus thrown away would amount, in the course of a tradesman's life, to a decent fortune for a child. Then there is the injury to health from these night adventures; there are the quarrels, there is the vicious habit of loose and filthy talk, there are the slanders and the backbitings, there are the admiration of contemptible wit, and there the scoffings at all that is sober and serious.'

The next even improves upon this :—

'Show your affection for your wife and your admiration of her not in nonsensical compliment, not in picking up her handkerchief or her glove, or in carrying her fan; not, though you have the means, in hanging trinkets and baubles upon her; not in making yourself a fool by winking at and seeming pleased with her foibles or follies or faults; but show them by acts of real goodness towards her; prove by unequivocal deeds the high value you set on her health and life and peace of mind; let your praise of her go to the full extent of her deserts, but let it be consistent with truth and with sense, and such as to convince her of your sincerity. He who is the flatterer of his wife only prepares her ears for the hyperbolical stuff of others. The kindest appellation that her Christian name affords is the best you can use, especially before faces. An everlasting "my dear" is but a sorry compensation for a want of that sort of love that makes the husband cheerfully toil by day, break his rest by night, endure all sorts of hardships, if the life or health of his wife demand it. Let your deeds and not your words carry to her heart a daily and hourly confirmation of the fact that you value her health and life and happiness beyond all other things in the world; and let this be manifest to her, particularly at those times when life is always more or less in danger.'

Cobbett left America in fierce wrath, after being prosecuted for a libel on Dr. Rush. His offence was marked; but his punishment for so free a country was, to say the least, not lenient. The case originated in his interference with the manner in which Dr. Rush treated his patients in the yellow fever. He accused him of Sangrado practice, or a too free use of the lancet; and it is amusingly characteristic of the witty and humourous malice of the man, to find him many years afterwards, when self-exiled to America, concluding a double-barrelled paragraph of his journal in these terms: 'An American counts the cost of powder and shot. If he is deliberate in everything else, this habit will hardly forsake him in the act of shooting. When the sentimental flesh-eaters hear the report of his gun, they may begin to pull out their white handkerchiefs; for death follows the pull of the trigger with perhaps even more certainty than it used to follow the lancet of Dr. Rush.'

A leading event in Cobbett's life was the severe fine and long imprisonment to which he was subjected, for daring to give way to the impulse which led him to denounce in warm, but only fitting terms, the flogging of Englishmen under the bayonets and sabres of Hanoverians. He was at this time living in the bosom of his family on his farm of Botley, in the midst of domestic enjoyment of no ordinary kind, and leading no inglorious or useless life. His long imprisonment, and the ruin of his affairs, left deep traces in a quick and resentful, but certainly not an ungenerous mind.

After a picture of domestic life which must charm everybody, and which is well worth the attentive study of every man and woman who has a family to train, he winds up:—

'In this happy state we lived until the year 1810, when the Government laid its hands upon me, dragged me from these delights, and crammed me into a jail amongst felons, of which I shall have to speak more fully, when, in the last number, I come to speak of the duties of the citizen. This added to the difficulties of my task of teaching; for now I was snatched away from the only scene in which it could, as I thought, properly be executed. But even these difficulties were got over. The blow was, to be sure, a terrible one; and how was it felt by these poor children? It was in the month of July when the horrible sentence was passed upon me. My wife, having left her children in the care of her good and affectionate sister, was in London waiting to know the doom of her husband. When the news arrived at Botley, the three boys — one eleven, another nine, and the other seven years old — were hoeing cabbages in that garden which had been the source of so much delight. When the account of the savage sentence was brought to them, the youngest could not for some time be made to understand what a jail was; and, when he did, he, all in a tremor, exclaimed, "Now, I'm sure, William, that papa is not in a place like that!" The other, in order to disguise his tears and smother his sobs, fell to work with the hoe, and chopped about like a blind person. This account, when it reached me, affected me more, filled me with deeper resentment, than any other circumstance. And, oh! how I despise the wretches who talk of my vindictiveness—of my exultation at the confusion of those who inflicted those sufferings! How I despise the base creatures, the crawling slaves, the callous and cowardly hypocrites, who affect to be "shocked" (tender souls!) at my expressions of joy at the death of Gibbs, Ellenborough, Percival, Liverpool, Canning, and the rest of the tribe that I have already seen out, and at the fatal workings of that system for endeavouring to check which I was thus punished!'

When the spy system had produced the horrors of 1817 and the Six Acts, Cobbett, who was still under heavy recognisances, thought it prudent for himself and his sureties to withdraw for a time to America. He imagined, not without cause, that one of the Six Acts was directly aimed at him; and the suspension of the Habeas Corpus Act made his situation very perilous. Cobbett therefore made the best of his way to Liverpool with his large young family; and from thence, upon the 26th March 1817, he addressed the public in these terms :—

'My departure for America will surprise nobody but those who do not reflect. A full and explicit statement of my reasons will appear in a few days, probably the 5th of April. In the meanwhile, I think it necessary for me to make known that I have fully empowered a person of respectability to manage and settle all my affairs in England. I owe my countrymen sincere regard, which I shall always entertain for them in a higher degree than towards any other people upon earth. I carry nothing from my country but my wife and my children, and surely they are my own at anyrate. I shall always love England better than any other country— I will never become a subject or citizen of any other state; but I and mine were not born under a government having the absolute power to imprison us at its pleasure, and, if we can avoid it, we will never live nor die under such an order of things. . . . When this order of things shall cease, then shall I again see England.'

By the disposal of his property at Botley, upon which he must have expended a great deal, and other transactions at this time, added to his ruinous imprisonment, law expenses, and a heavy fine of a thousand pounds which had been imposed upon him, his pecuniary affairs suffered serious derangement, from which they probably never recovered.

In America he took a farm, or at least a house in the country with some land, resumed his indefatigable habits, and opened a seed store in New York. The *Registers* came regularly across the Atlantic, and were eagerly expected. Another of Cobbett's books, the *Year's Residence in America*, now appeared in parts.

Cobbett returned to England as soon as the Habeas Corpus Suspension Bill had expired, and, settling at Kensington, recommenced his labours as a journalist. These were, indeed, never suspended, save while he was at sea.

In the autumn of 1822 he began his *Rural Rides*, which he continued for five different seasons, and in which he indulged his natural love for rural objects, and everything connected with country life. He seems to have had a true and lively feeling for the beautiful in nature, and the pure and simple taste which is ever the attendant of this kind of sensibility. He always travelled on horseback, accompanied by one or other of his sons, and showed his good taste by departing from the usual thoroughfares, and finding his way across fields, by footpaths, by-lanes, bridle-ways, and hunting-gates—'steering' over the country, as he expresses it, for such landmarks as village spires and old chapels. His object was to see and converse with the farmers and labourers in their own abodes, to look at the crops, to survey the modes of husbandry. The 'agricultural interest' was beginning to suffer smartly by this time, and the gridiron was adorning every number of the *Register*. Politics mingle largely in the journal of the *Rural Rides*, but only to increase their vivacity and render them more piquant; and, when Cobbett leaves Bolt Court and rides abroad to air his notions, he always becomes mellow in spirit—gay, and good-humoured.

The remainder of Cobbett's career, which was so full of inconsistencies, may be briefly summed up. On his return to England in 1800, he had published the *Porcupine* and *Weekly Register*, the latter of which was continued up till the time of his death. It appeared at first as a Tory, but became eventually a Radical publication. It abounded in violent personal and political attacks on public men. He was, as has already been noted, twice fined and prosecuted for libel; and in 1809, for the publication of a libel relating to the flogging of some men in the local militia at Ely, he had been sentenced to two years' imprisonment in Newgate, to pay £1000 to the king, and on his release to give security for his good behaviour for seven years, himself in £300, and two securities in £100 each. As already mentioned, he went to America, and returned in 1819. Two separate attempts made to enter Parliament in 1820 and 1826 both failed. In 1831 he was again tried for libel, when he acquitted himself with a memorable speech, and the jury being equally divided on the case, he was discharged. In 1833 he entered Parliament as member for Oldham, but found the late hours and stifling atmosphere of Parliamentary life unsuited to his simple tastes. His life, which had been one of unceasing literary industry, was brought to a close by an attack of disease of the throat, from which he never recovered. He died 17th June 1835. His writings, which deal with rural life, have been commended as having been widely and practically useful. Besides his political writings, including twenty volumes of *Parliamentary Debates*, etc., Cobbett wrote his *Cottage Economy*, *English Grammar*, *History of the Protestant Reformation*, and *Rural Rides*, etc. His language, as will be seen from the extracts we have given, is uniformly forcible and vigorous, and, as he himself says, 'his popularity' was owing to his 'giving truth in clear language.'

The announcement of the death of Cobbett's eldest daughter appeared in the *Times* of October 26, 1877. She was born in Philadelphia in 1795, where her father was residing. She died at Brompton Crescent, London, in her eighty-second year. In 1810–12, while her father was imprisoned in Newgate for libel, she kept him company, acting as his amanuensis and the custodian of his papers, and writing at his dictation leading articles for his weekly publication. Some of Cobbett's most stirring articles are said to have been sent to press in the handwriting of Miss Cobbett.

Another member of Cobbett's family was M.P. for Oldham; and his eldest son, the eccentric William Cobbett, died suddenly on January 12, 1878, in one of the central halls of the Houses of Parliament, whither he had come to further a measure of litigation.

HUGH MILLER.

THE name of Hugh Miller is one which commands universal regard and respect, whether we view him as a geologist, a man of letters, or as a stone-mason, who possessed sturdy independence of character and indomitable perseverance. In telling the story of his life, we have at least two good sources of information. There is the interesting autobiography which he wrote, *My Schools and Schoolmasters; or, The Story of my Education*, and also *The Life and Letters of Hugh Miller*, by Peter Bayne, LL.D.

Hugh Miller was born in the town of Cromarty, 10th October 1802. His father, who was brave and gentle, and seldom angry without just cause, had a strange dream regarding his first-born. There was a dash of Celtic blood in his descent, but his character belonged more to the lowland type. His paternal ancestors had all been seafaring men; and for more than a hundred years before his birth not one of these ancestors had been laid to rest in the churchyard of Cromarty. His own father perished at sea when he was but five years old. Of this sad event, and before the news of it had arrived at his home, Miller writes:—

'There were no forebodings in the master's dwelling; for

his Peterhead letter—a brief but hopeful missive—had been just received; and my mother was sitting, on the evening after, beside the household fire, plying the cheerful needle, when the house door, which had been left unfastened, fell open, and I was despatched from her side to shut it. What follows must be regarded as simply the recollection, though a very vivid one, of a boy who had completed his fifth year only a month before. Day had not wholly disappeared, but it was fast posting on to night, and a grey haze spread a neutral tint of dimness over every more distant object, but left the nearer ones comparatively distinct, when I saw at the open door, within less than a yard of my breast, as plainly as ever I saw anything, a dissevered hand and arm stretched towards me. Hand and arm were apparently those of a female: they bore a livid and sodden appearance; and directly fronting me, where the body ought to have been, there was only blank, transparent space, through which I could see the dim forms of the objects beyond. I was fearfully startled, and ran shrieking to my mother, telling what I had seen; and the house girl whom she next sent to shut the door, apparently affected by my terror, also returned frightened, and said that she too had seen the woman's hand; which, however, did not seem to be the case. And finally, my mother going to the door, saw nothing, though she appeared much impressed by the extremeness of my terror and the minuteness of my description. I communicate the story, as it lies fixed in my memory, without attempting to explain it. The supposed apparition may have been merely a momentary affection of the eye, of the nature described by Sir Walter Scott in his *Demonology*, and Sir David Brewster in his *Natural Magic*. But if so, the affection was one of which I experienced no after return; and its coincidence, in the case, with the probable time of my father's death, seems at least curious.'

This superstitious feeling was no doubt nursed by his mother, who entertained a belief in fairies, witches, dreams, ghosts, and presentiments. She was but eighteen when married, while her husband was forty-four. Young Hugh was sent to a dame school, where he learned to read, and during his sixth year spelt through the Shorter Catechism, the Proverbs, and the New Testament. He read the Old Testament narrative, especially the story of Joseph, with growing interest. He also perused those classics for youth, *Jack the Giant-Killer*, *Jack and the Bean Stalk*, and followed them up with Pope's Homer's *Iliad* and *Odyssey*, and Bunyan's *Pilgrim's Progress*. In process of time he also devoured all the voyages, travels, and romances upon which he could lay his hands. His mother, a young widow with her son of five, and two daughters emerging from infancy, with a fixed income of but twelve pounds, betook herself to her needle, and was otherwise befriended by her two brothers, mentioned in *My Schools and Schoolmasters* under the names of Uncle James and Uncle Sandy. Thinking themselves called upon to take his father's place in the work of his instruction and discipline, Miller remarks that he owed much more of his real education to them than to any of the teachers whose schools he afterwards attended.

'My elder uncle, James,' he writes, 'added to a clear head and much native sagacity, a singularly retentive memory, and great thirst of information. He was a harness-maker, and wrought for the farmers of an extensive district of country; and as he never engaged either journeyman or apprentice, but executed all his work with his own hands, his hours of labour, save that he indulged in a brief pause as the twilight came on, and took a mile's walk or so, were usually protracted from six o'clock in the morning till ten at night. Such incessant occupation left him little time for reading; but he often found some one to read beside him during the day; and in the winter evenings his portable bench used to be brought from his shop at the other end of the dwelling, into the family sitting-room, and placed beside the circle round the hearth, where his brother Alexander, my younger uncle, whose occupation left his evenings free, would read aloud from some interesting volume for the general benefit, placing himself always at the opposite side of the bench, so as to share in the light of the worker. Occasionally the family circle would be widened by the accession of from two to three intelligent neighbours, who would drop in to listen; and then the book, after a space, would be laid aside, in order that its contents might be discussed in conversation. In the summer months, Uncle James always spent some time in the country in looking after and keeping in repair the harness of the farmers for whom he wrought; and during his journeys and twilight walks on these occasions, there was not an old castle, or hill fort, or ancient encampment, or antique ecclesiastical edifice, within twenty miles of the town, which he had not visited and examined over and over again. He was a keen local antiquary, knew a good deal about the architectural styles of the various ages at a time when these subjects were little studied or known, and possessed more traditionary lore, picked up chiefly in his country journeys, than any man I ever knew. What he once heard he never forgot, and the knowledge which he had acquired he could communicate pleasingly and succinctly, in a style which, had he been a writer of books, instead of merely a reader of them, would have had the merit of being clear and terse, and more laden with meaning than words. From his reputation for sagacity, his advice used to be much sought after by the neighbours in every little difficulty that came their way; and the counsel given was always shrewd and honest. I never knew a man more entirely just in his dealings than Uncle James, or who regarded every species of meanness with a more thorough contempt. I soon learned to bring my story-books to his workshop, and became, in a small way, one of his *readers*—greatly more, however, as may be supposed, on my own account than his. My books were not yet of the kind which he would have chosen for himself; but he took an interest in *my* interest; and his explanations of all the hard words saved me the trouble of turning over a dictionary. And when tired of reading, I never failed to find rare delight in his anecdotes and old-world stories, many of which were not to be found in books, and all of which, without apparent effort on his own part, he could render singularly amusing. Of these narratives, the larger part died with him; but a portion of them I succeeded in preserving in a little traditionary work published a few years after his death. I was much a favourite with Uncle James—even more, I am disposed to think, on my father's account than on that of his sister, my mother. My father and he had been close friends for years, and in the vigorous and energetic sailor he had found his beau-ideal of a man.

'My Uncle Alexander was of a different cast from his brother, both in intellect and temperament; but he was characterised by the same strict integrity; and his religious feelings, though quiet and unobtrusive, were perhaps more deep. James was somewhat of a humorist, and fond of a good joke. Alexander was grave and serious, and never, save on one solitary occasion, did I know him even attempt a jest. On hearing an intelligent but somewhat eccentric neighbour observe, that "all flesh is grass," in a strictly physical sense, seeing that all the flesh of the herbivorous animals is elaborated from vegetation, and all the flesh of the carnivorous animals from that of the herbivorous ones, Uncle Sandy remarked that, knowing as he did the piscivorous habits of the Cromarty folk, he should surely make an exception in his generalization, by admitting that in at least one village "all flesh is fish." My uncle had acquired the trade of the cartwright, and was employed in a workshop at Glasgow at the time the first war of the French Revolution broke out, when, moved by some such spirit as possessed his uncle (the victim of Admiral Vernon's unlucky expedition) or Old Donald Roy, when he buckled himself to his Highland broadsword, and set out in pursuit of the caterans, he entered the navy. . . .

'Early on the Sabbath evenings I used regularly to attend at my uncle's with two of my maternal cousins, boys of about my own age, and latterly with my two sisters, to be catechized, first on the Shorter Catechism, and then on the Mother's Catechism of Willison. On Willison my uncles always cross-examined us, to make sure that we understood the short and simple questions; but, apparently regarding the questions of the Shorter Catechism as seed sown for a future day, they were content with having them well fixed in our memories. There was a Sabbath class taught in the parish church at the time by one of the elders; but Sabbath schools my uncles regarded as merely compensatory institutions, highly creditable to the teachers, but very discreditable indeed to the parents and relatives of the taught; and so they of course never thought of sending us there. Later in the evening, after a short twilight walk, for which the sedentary occupation of my Uncle James formed an apology, but in which my Uncle Alexander always shared, and which usually led them into solitary woods, or along an unfrequented sea-shore, some of the old divines were read; and I used to take my place in the circle, though, I am afraid, not to much advantage. I occasionally caught a fact, or had my attention arrested for a moment by a simile or metaphor; but the trains of close argument, and the passages of dreary "application," were always lost.

'I quitted the dame's school at the end of the first twelvemonth, after mastering that grand acquirement of my life—the art of holding converse with books, and was transferred straightforth to the grammar school of the parish, at which there attended at this time about a hundred and twenty boys, with a class of about thirty individuals more, much looked down upon by the others, and not deemed greatly worth the counting, seeing that it consisted only of *lassies*. . . . The building in which we met was a low, long, straw-thatched cottage, open from gable to gable, with a mud floor below and an unlathed roof above; and stretching along the naked rafters, which, when the master chanced to be absent for a few minutes, gave noble exercise in climbing, there used frequently to lie a helm, or oar, or boathook, or even a foresail, the spoil of some hapless peat-boat from the opposite side of the firth. . . . The parish schoolmaster was a scholar and an honest man, and if a boy really wished to learn, *he* certainly could teach him. . . . He was in the habit of advising the parents or relations of those he deemed his clever lads, to give them a classical education; and meeting one day with Uncle James, he urged that I should be put on

Latin. I was a great reader, he said; and he found that when I missed a word in my English tasks, I almost always submitted a synonym in the place of it. And so, as Uncle James had arrived, on data of his own, at a similar conclusion, I was transferred from the English to the Latin form, and, with four other boys, fairly entered on the *Rudiments*. I laboured with tolerable diligence for a day or two; but there was no one to tell me what the rules meant, or whether they really meant anything; and when I got on as far as *penna*, a pen, and saw how the changes were rung on one poor word, that did not seem to be of more importance in the old language than in the modern one, I began miserably to flag, and to long for my English reading, with its nice amusing stories and its picture-like descriptions. The *Rudiments* was by far the dullest book I had ever seen. It embodied no thought that I could perceive: it certainly contained no narrative; it was a perfect contrast to not only the *Life and Adventures of Sir William Wallace*, but to even the voyages of Cook and Anson. None of my class-fellows were by any means bright: they had been all set on Latin without advice of the master; and yet, when he learned, which he soon did, to distinguish and call us up to our tasks by the name of the " heavy class," I was, in most instances, to be found at its nether end. Shortly after, however, when we got a little further on, it was seen that I had a decided turn for translation. The master, good simple man that he was, always read to us in English, as the school met, the piece of Latin given us as our task for the day; and as my memory was strong enough to carry away the whole translation in its order, I used to give him back in the evening, word for word, his own rendering, which satisfied him on most occasions tolerably well. There were none of us much looked after; and I soon learned to bring books of amusement to the school with me, which, amid the Babel confusion of the place, I contrived to read undetected. Some of them, save in the language in which they were written, were identical with the books proper to the place. I remember perusing by stealth in this way Dryden's *Virgil*, and the *Ovid* of Dryden and his friends; while Ovid's own *Ovid* and Virgil's own *Virgil* lay beside me, sealed up in the fine old tongue, which I was thus throwing away my only chance of acquiring.

'One morning, having the master's English rendering of the day's task well fixed in my memory, and no book of amusement to read, I began gossiping with my nearest class-fellow, a very tall boy, who ultimately shot up into a lad of six feet four, and who on most occasions sat beside me, as lowest in the form save one. I told him about the tall Wallace and his exploits, and so effectually succeeded in awakening his curiosity that I had to communicate to him, from beginning to end, every adventure recorded by the blind minstrel. My story-telling vocation once fairly ascertained, there was, I found, no stopping in my course. I had to tell all the stories I ever heard or read—all my father's adventures, so far as I knew them, and all my Uncle Sandy's, with the story of Gulliver, and Philip Quarll, and Robinson Crusoe—of Sinbad, and Ulysses, and Mrs. Radcliffe's heroine Emily, with, of course, the love passages left out; and at length, after weeks and months of narrative, I found my available stock of acquired fact and fiction fairly exhausted. The demand on the part of my class-fellows was, however, as great and urgent as ever, and, setting myself, in the extremity of the case, to try my ability of original production, I began to dole out to them, by the hour and the diet, long *extempore* biographies, which proved wonderfully popular and successful. My heroes were usually warriors like Wallace, and voyagers like Gulliver, and dwellers in desolate islands like Robinson Crusoe; and they had not unfrequently to seek shelter in huge deserted castles, abounding in trap-doors and secret passages like that of Udolpho. . . . With all my carelessness, I continued to be a sort of favourite with the master, and, when at the general English lesson, he used to address to me little quiet speeches, vouchsafed to no other pupil, indicative of a certain literary ground common to us, on which the others had not entered. " That, sir," he has said, after the class had just perused, in the second collection, a *Tatler* or *Spectator*—" that, sir, is a good paper: it's an *Addison;*" or, " That's one of Steele's, sir;" and on finding in my copy-book, on one occasion, a page filled with rhymes, which I had headed " Poem on Care," he brought it to his desk, and, after reading it carefully over, called me up, and with his closed penknife, which served as a pointer, in the one hand, and the copy-book brought down to the level of my eyes in the other, began his criticism. " That's bad grammar, sir," he said, resting his knife-handle on one of the lines; " and here's an ill-spent word; and there's another; and you have not at all attended to the punctuation; but the general sense of the piece is good—very good indeed, sir." And then he added, with a grim smile, " *Care*, sir, is, I daresay, as you remark, a very bad thing; but you may safely bestow a little more of it on your spelling and your grammar."

'There were several other branches of my education going on at this time outside the pale of the school, in which, though I succeeded in amusing myself, I was no trifler. The shores of Cromarty are strewed over with water-rolled fragments of the primary rocks, derived chiefly from the west during the ages of the boulder clay; and I soon learned to take a deep interest in sauntering over the various pebble beds when shaken up by recent storms, and in learning to distinguish their numerous components. But I was sadly in want of a vocabulary; and as, according to Cowper, "the growth of what is excellent is slow," it was not until long after that I bethought me of the obvious enough expedient of representing the various species of simple rocks by certain numerals, and the compound ones by the numerals representative of each separate component, ranged, as in vulgar fractions, along a medial line, with the figures representative of the prevailing materials of the mass above, and those representative of the materials in less proportions below. Though, however, wholly deficient in the signs proper to represent what I knew, I soon acquired a considerable quickness of eye in distinguishing the various kinds of rock, and tolerably definite conceptions of the generic character of the porphyries, granites, gneisses, quartz rocks, clay slates, and mica schists which everywhere strewed the beach. In the rocks of mechanical origin I was at this time much less interested; but in individual, as in general history, mineralogy almost always precedes geology.'

When twelve years of age, Miller wrote some verses on a singular adventure which he and another companion experienced in a place called the Doocot Cave. He wrote four successive accounts of this experience. The first was executed in enormously bad verse, which, however, excited the wonder of Miss Bond, the mistress of the Cromarty boarding-school; at nineteen he altered and polished the verses; in the vigour of early manhood he described the adventure in a letter to Principal Baird; and when over fifty years of age, he gave the most glowing and perfect account of all in *Schools and Schoolmasters.*

After this romantic occurrence, he visited some friends in the highlands of Sutherland, which had the effect of familiarizing his mind with the scenery, character, and condition of the Highlands. At this time he showed considerable activity of mind in the peculiar character of his amusements. He made small vessels like those he had read about in the voyages of Anson and Cook, and launched them in a horse-pond. In turn he tried chemistry, and painting, and sculpture, and palmistry. He would also draw a map of a particular country in the sand, and, having collected quantities of variously-coloured shells from the beach, he arranged them in such a way as to represent its inhabitants; or, heading a band of school-fellows, they would penetrate one of the steepest precipices on the south foot of the hill of Cromarty, and personate outlaws and buccaneers. Meanwhile his mother and uncles found him a troublesome lad to manage. He would sometimes play truant from school for three weeks out of four, and he continued obstinate and wilful. In the winter of 1816, he lost both his two little sisters, and could not but be touched at his mother's grief. His schooling finished when he was fifteen in a pitched battle with his teacher. Before that time he had proved himself a desperate fighter in his combats with the other boys. In his fight with his master, he was mauled in a way that filled him with aches and bruises for a full month after.

After a widowhood of more than thirteen years, his mother married again, and he was forced to begin and work in earnest at some trade; so he determined being a mason. Without objecting to the match, 'you may be certain,' he wrote to a friend some years later, 'that it gave me much disgust at the time.' In making this decision, he thought that perhaps literature or natural science might be his proper vocation, but at the same time he determined that much of his leisure, in spite of his misspent youth, should be given to the study of the best English authors. His Uncle James would have liked had he chosen some of the learned professions requiring a college training, such as a lawyer or a minister. But as they were all decided that a minister could not be manufactured by a few years' study, they at length consented that he should begin a life of manual labour. He was accordingly apprenticed to the husband of one of his maternal aunts for a term of three years, and he began work in earnest. 'Noble, upright, self-relying Toil!' he writes, 'who that knows thy solid worth and value would be ashamed of thy hard hands, and thy soiled vestments, and thy obscure tasks—thy humble cottage, and hard couch, and homely fare! Save for thee and thy lessons, man in society would everywhere sink into a sad compound of the fiend and the wild beast; and this fallen world would be as certainly a moral as a natural wilderness. But I little thought of the excellence of thy character and of thy teachings, when, with a heavy heart, I set out about this time, on a morning of early spring, to take my first lesson from thee in a sandstone quarry.' The work oppressed his growing frame at first, but use gave ease to the young stonemason. How he escaped the vice of 'dram-drinking' is thus related by himself:—'The drinking usages of the profession in which I laboured were at this time many; when a foundation was laid, the workmen were treated to drink; they were treated to drink when the walls were levelled for laying the joists; they were treated to drink when the building was finished; they were treated to drink when an apprentice joined the squad; treated to drink when his "apron was washed;" treated to drink when "his time was out;" and occasionally they learned to treat one another to drink. In laying down the foundation-stone of one of the larger houses built this year by Uncle David and his partner, the workmen had a royal "founding-pint," and two whole glasses of the whisky came to my share. A full-grown man would not have deemed a gill of usquebaugh an overdose, but it was considerably too much for me; and when the party broke up, and I got home to my books, I found, as I opened the pages of a favourite author, the letters dancing before my eyes, and that I could no longer master the sense. I have the volume at present before me—a small edition of the essays of Bacon, a good deal worn at the corners by the friction of the pocket; for of Bacon I never tired. The condition into which I had brought myself was, I felt, one of degradation. I had sunk, by my own act, for the time, to a lower level of intelligence than that on which it was my privilege to be placed; and though the state could have been no very favourable one for forming a resolution, I in that hour determined that I should never again sacrifice my capacity of intellectual enjoyment to a drinking usage; and, with God's help, I was enabled to hold by the determination.'

During the winter, when mason work was no longer possible, he paid a visit to Strathcarron, a wild Highland glen, where he made some observations on a Scotch pine forest. His year of toil had rendered him sober and thoughtful. He formed a bosom friend in a young house-painter called William Ross. 'He was a lad of genius,' writes Miller, 'drew truthfully, had a nice sense of the beautiful, and possessed the true poetic faculty; but he lacked health and spirits, and was naturally of a melancholy temperament, and diffident of himself.' This friendship was of the highest importance to Miller. Ross told him that his drawings and verses were but commonplace, that he would be better employed in cultivating his writing powers, in turning his attention to the cultivation of a good prose style. In the spring of 1821, work was again resumed. He laboured for a while at Cononside, and was introduced for the first time to the barrack or bothy life amongst a squad of rough masons. Amid this barbarous life he was not entirely unhappy. Their food consisted for the most part of oatmeal porridge or cakes, and milk when it could be got. He was charmed with the scenery of Strathpeffer, about five miles from Cononside; and during the summer nights, when he had from three to four hours to himself, he could explore the valleys and climb the ridges of the hills with which he was surrounded. He dived into the woods and feasted on the raspberries and gueans to be found there. 'My recollections,' he said afterwards, 'of this rich tract of country, with its woods and towers, and noble river, seem as if bathed in the red light of gorgeous sunsets.' In a letter to William Ross, he thus spoke of this period: 'When the task of the day was over, and I walked out amid the fields and woods to enjoy the cool of the evening, it was then that I was truly happy. Before me the Conon rolled her broad stream to the sea; behind, I seemed shut up from all intercourse with mankind by a thick and gloomy wood; while the tower of Fairburn, and the blue hills behind it, formed the distant landscape. Not a cloud rose upon the sky, not a salmon glided beneath me in the river, nor a leaf shook upon the alders that o'erhung the stream, but raised some poetic emotion in my breast.' The next winter and spring were spent at Cromarty, where he again met William Ross, but in the working season of 1822 he returned to Cononside. In a letter written at this period, he says: 'I had determined early this season to conform to every practice of the barrack, and as I was an apt pupil, I had in a short time become one of the freest, and not the least rude of its inmates. I became an excellent baker, and one of the most skilful of cooks. I made wonderful advances in the art of practical joking, and my *bon-mots* were laughed at and repeated. There were none of my companions who could foil me in wrestling, or who could leap within a foot of me; and after having taken the slight liberty of knocking down a young fellow who insulted me, they all began to esteem me as a lad of spirit and promise.'

The neighbourhood of Cononside enabled Miller to extend his geologic explorations in a definite direction. He had been urged by the foreman of the squad with which he was connected to study geometry and architecture, and these he pursued for some time. He finished his apprenticeship on the 11th November 1822, amidst great hardship, while working at a wall and farm-steading in the neighbourhood of Cromarty, often standing 'day after day with wet feet in a water-logged ditch.' 'How these poor hands of mine,' he says, 'burnt and beat at night at this time, as if an unhappy heart had been stationed in every finger; and what cold chills used to run, sudden as electric shocks, through the feverish frame.' Ere the winter was over, he had gained his ordinary robust health.

'I read, wrote, drew,' he says, 'corresponded with my friend William Ross (who had removed to Edinburgh), re-examined the Eathie Lias, and re-explored the Eathie Burn—a noble old red sandstone ravine, remarkable for the wild picturesqueness of its cliffs and the beauty of its cataracts. I spent, too, many an evening in Uncle James's workshop, on better terms with both my uncles than almost ever before—a consequence, in part, of the sober complexion which, as the seasons passed, my mind was gradually assuming, and in part, of the manner in which I had completed my engagement with my master. "Act always," said Uncle James, "as you have done in this matter. In all your dealings, give your neighbour the *cast of the bauk*—'good measure, heaped up and running over'—and you will not lose by it in the end." I certainly did not lose by faithfully serving out my term of apprenticeship. It is not uninstructive to observe how strangely the public are led at times to attach paramount importance to what is in reality only subordinately important, and to pass over the really paramount without thought or notice. The destiny in life of the skilled mechanic is much more influenced, for instance, by his second education—that of his apprenticeship—than by his first—that of the school; and yet it is to the education of the school that the importance is generally regarded as attaching, and we never hear of the other. The careless, incompetent scholar has many opportunities of recovering himself; the careless, incompetent apprentice, who either fails to serve out his regular time, or who, though he fulfils his term, is discharged an inferior workman, has very few; and further, nothing can be more certain than that inferiority as a workman bears much more disastrously on the condition of the mechanic than inferiority as a scholar. Unable to maintain his place among brother journeymen, or to render himself worthy of the average wages of his craft, the ill-taught mechanic falls out of regular employment, subsists precariously for a time on occasional jobs, and either, forming idle habits, becomes a vagabond *tramper*, or, getting into the toils of some rapacious taskmaster, becomes an enslaved *sweater*. For one workman injured by neglect of his school education, there are scores ruined by neglect of their apprenticeship education. Three-fourths of the distress of the country's mechanics (of course not reckoning that of the unhappy class who have to compete with machinery), and nine-tenths of their vagabondism, will be found restricted to inferior workmen, who, like Hogarth's "careless apprentice," neglected the opportunities of their second term of education. The sagacious painter had a truer insight into this matter than most of our modern educationists.'

Miller's first kindly act on becoming a journeyman was to build a cottage for his Aunt Jenny, on a piece of ground he had inherited from his father in Cromarty. The cottage still stands a worthy monument of such an act. In his correspondence with William Ross, who was now in Edinburgh, he enclosed from time to time a selection from his poems. These poems, says Mr. Bayne, 'are fluent and vivacious, but display little original power or depth of melody.' In midsummer 1823, he found work at Gairloch, on the west coast of Ross-shire. He was now not yet twenty-one, and the different hovels he lived in were damp and uncomfortable, and his food was of the plainest, often oatmeal without milk. The winter of 1823 he spent as usual at Cromarty, but a small property at Leith having been left to his mother, which had been a constant source of annoyance ever since his father's death, he left in the spring of 1824 to investigate the affair on the spot. He had also determined to try his fortune among the stonecutters of Edinburgh, 'perhaps the most skilful in their profession in the world.' On the fourth day after leaving Cromarty, his vessel was threading the waters of the Firth of Forth. 'Many a long-cherished association drew my thoughts to Edinburgh. I was acquainted with Ramsay, and Fergusson, and the " Humphrey Clinker" of Smollett, and had read a description of the place in the *Marmion* and the earlier novels of Scott, and I was not yet too old to feel as if I were approaching a great magical city—like some of those in the *Arabian Nights* —that was even more intensely poetical than nature itself. I did somewhat chide the tantalizing mist, that, like a capricious showman, now raised one corner of its curtain, and anon another, and showed me the place at once very indistinctly, and only by bits at a time; and yet I know not that I could in reality have seen it to greater advantage, or after a mode more in harmony with my previous conceptions. The water in the harbour was too low during the first hour or two after our arrival to float our vessel, and we remained tacking in the roadstead, watching for the signal from the pier-head, which was to intimate to us when the tide had risen high enough for our admission; and so I had sufficient time given me to con over the features of the scene, as presented in detail. At one time a flat reach of the New Town came full into view, along which, in the general dimness, the multitudinous chimneys stood up like stacks of corn in a field newly reaped; at another, the castle loomed out dark in the cloud; then, as if suspended over the earth, the rugged summit of Arthur's Seat came strongly out, while its base still remained invisible in the wreath; and anon I caught a glimpse of the distant Pentlands, enveloped by a clear blue sky, and lighted up by the sun.

Leith, with its thicket of masts, and its tall round tower, lay deep in shade in the foreground—a cold, dingy, ragged town, but so strongly relieved against the pale smoky grey of the background, that it seemed another little city of Zoar, entire in front of the burning.'

Hugh Miller visited the burying-grounds, the churches, and the various places of historical interest in and around Edinburgh, and found employment at his trade at Niddrie House, in the neighbourhood. That his life there ministered to his growth is abundantly evident, although the workmen with whom he was obliged to associate were many of them of a low type of character. After working two seasons at Niddrie, he returned to Cromarty, where he was welcomed by his uncle, his cousin George, and other relatives. His two years' work had given him the 'stonecutter's malady,' which probably weakened his lungs for life. At home he renewed acquaintanceship with John Swanson, and corresponded with William Ross, whom he had left in Edinburgh. His friendship and correspondence with Swanson was of immense benefit to him from a religious point of view. Swanson would not let young Miller rest until he had asked of himself a reason for the faith that was in him. In January 1826 he wrote to this effect: 'Go on, my dear Hugh, go on, and the Lord Himself will bless you. If you are not under the teaching of the Spirit of God I am deceived, and if I do not find you soon established in the way of happiness, peace, and life, I shall be miserably disappointed.' Miller in turn, in writing to his friend Ross, assumed the same tone of friendly earnestness regarding his highest welfare. He had in the meantime pursued his occupation as stonecutter, as health and opportunity permitted. In the spring of 1828 he drew up a list, headed, 'Things which I intend doing, but many of which, experience says, shall never be done.' This list comprised many projects in geometry, architecture, sculpture, and in literary prose composition. In the summer he formed the acquaintanceship of Dr. (then Mr.) Robert Carruthers, editor of the *Inverness Courier*, which was afterwards of so much use to him. Dr. Carruthers printed for him a collection of his verses, under the title of *Poems written in the Leisure Hours of a Journeyman Mason*, and his book met with a moderate success. Amongst the friends which the publication of his poems procured for him was a Mr. Strahan, who also wrote poetry. One of his sons was Alexander Strahan, whom Miller introduced to Messrs. Johnstone & Hunter, publishers, Edinburgh, with a view of learning the business, and who has since become well known in the publishing world.

After many experiments in versification, he made up his mind that poetry was not his proper vocation, and accordingly he next tried prose. His famous letters on the herring fishing were written for the *Inverness Courier* in the summer of 1829, and attracted much attention. The publication of his poems, though anonymous, yet made him the literary lion of Cromarty, and extended his friendships even beyond. Sir Thomas Dick Lauder, Mr. Isaac Forsyth of Elgin, Miss Dunbar of Boath, and Principal Baird, for whom he wrote a sketch of his life up till 1825, all gave him their patronage in the disposal of his volume. In 1831 he was vitally concerned in a newspaper correspondence. The minister of the Gaelic Chapel, Cromarty, had petitioned to the effect that he either be assigned a parish within the bounds of the parish of Cromarty, or a collegiate charge with Rev. Mr. Stewart. Hugh Miller acted as the mouthpiece of nearly eight hundred of his fellow-parishioners, and the letters in the newspapers and the pamphlet he published were successful in averting the possibility of this proposal becoming a reality.

While working in the churchyard at his occupation of stone-cutting, he had occasional visitors. His own minister would come and chat with him for hours together, and he also saw Sir Thomas Dick Lauder and Professor Pillans while thus engaged. In the summer of 1831 he first saw his future wife, Miss Lydia Mackenzie Fraser. Talking with two ladies beside a sun-dial in his uncle's garden, she 'came hurriedly tripping down the garden walk' and joined them. She was in her nineteenth year at the time, and, as described by Miller, 'she was very pretty, with a light *petite* figure, waxen clearness of complexion, making her look more like a fair child than a grown woman.'

The growth of the intimacy with Miss Fraser is thus pleasantly told in Miller's autobiography :—

'Only a few evenings after, I met the same young lady, in circumstances of which the writer of a tale might have made a little more. I was sauntering, just as the sun was sinking, along one of my favourite walks on the hill — a tree-skirted glade, now looking out through the openings on the ever-fresh beauties of the Cromarty Firth, with its promontories, and bays, and long lines of winding shore, and anon marking how redly the slant light fell through interstitial gaps on pale lichened trunks and huge boughs, in the deeper recesses of the wood, when I found myself unexpectedly in the presence of the young lady of the previous evening. She was sauntering through the wood as leisurely as myself, now and then dipping into a rather bulky volume which she carried, that had not in the least the look of a novel, and which, as I subsequently ascertained, was an elaborate essay on causation. We, of course, passed each other on our several ways without sign of recognition. Quickening her pace, however, she was soon out of sight; and I just thought, on one or two occasions afterwards, of the apparition that had been presented as she passed, as much in keeping with the adjuncts—the picturesque forest and the gorgeous sunset. It would not be easy, I thought, were the large book but away, to furnish a very lovely scene with a more suitable figure. Shortly after, I began to meet the young lady at the charming tea-parties of the place. Her father, a worthy man, who, from unfortunate speculations in business, had met with severe losses, was at this time several years dead; and his widow had come to reside in Cromarty, on a somewhat limited income, derived from property of her own. Liberally assisted, however, by relations in England, she had been enabled to send her daughter to Edinburgh, where the young lady received all the advantages which a first-rate education could confer. By some lucky chance, she was there boarded, with a few other ladies, in early womanhood, in the family of Mr. George Thomson, the well-known correspondent of Burns, and passed under his roof some of her happiest years. Mr. Thomson—himself an enthusiast in art—strove to inoculate the youthful inmates of his house with the same fervour, and to develop whatever seeds of taste or genius could be found in them; and, characterized till the close of a life extended far beyond the ordinary term by the fine chivalrous manners of the thorough gentleman of the old school, his influence over his young friends was very great, and his endeavours, in at least some of the instances, very successful. And in none, perhaps, was he more so than in the case of the young lady of my narrative. From Edinburgh she went to reside with the friends in

England to whose kindness she had been so largely indebted; and with them she might have permanently remained, to enjoy the advantages of superior position. She was at an age, however, which rarely occupies itself in adjusting the balance of temporal advantage; and her only brother having been admitted, through the interest of her friends, as a pupil into Christ's Hospital, she preferred returning to her widowed mother, left solitary in consequence, though with the prospect of being obliged to add to her resources by taking a few of the children of the town as day pupils.

'Her claim to take her place in the intellectual circle of the burgh was soon recognised. I found that, misled by the extreme youthfulness of her appearance, and a marked juvenility of manner, I had greatly mistaken the young lady. That she should be accomplished in the ordinary sense of the term—that she should draw, play, and sing well—would be what I should have expected; but I was not prepared to find that, mere girl as she seemed, she should have a decided turn, not for the lighter, but for the severer walks of literature, and should have already acquired the ability of giving expression to her thoughts in a style formed on the best English models, and not in the least like that of a young lady. The original shyness wore away, and we became great friends. I was nearly ten years her senior, and had read a great many more books than she; and, finding me a sort of dictionary of fact, ready of access, and with explanatory notes attached, that became long or short just as she pleased to draw them out by her queries, she had, in the course of her amateur studies, frequent occasion to consult me. There were, she saw, several ladies of her acquaintance, who used occasionally to converse with me in the churchyard; but in order to make assurance doubly sure respecting the perfect propriety of such a proceeding on her part, she took the laudable precaution of stating the case to her mother's landlord, a thoroughly sensible man, one of the magistrates of the burgh, and an elder of the kirk; and he at once certified that there was no lady of the place who might not converse, without remark, as often and as long as she pleased with me. And so, fully justified, both by the example of her friends—all very judicious women, some of them only a few years older than herself—and by the deliberate judgment of a very sensible man, the magistrate and elder, my young lady friend learned to visit me in the churchyard, just like the other ladies, and, latterly at least, considerably oftener than any of them. We used to converse on all manner of subjects connected with the *belles lettres* and the philosophy of mind, with, so far as I can at present remember, only one marked exception. On that mysterious affection which sometimes springs up between persons of the opposite sexes when thrown much together,—though occasionally discussed by the metaphysicians, and much sung by the poets,—we by no chance ever touched. Love formed the one solitary subject which, from some curious contingency, invariably escaped us.

'And yet, latterly at least, I had begun to think about it a good deal. Nature had not fashioned me one of the sort of people who fall in love at first sight. I had even made up my mind to live a bachelor life, without being very much impressed by the magnitude of the sacrifice; but I daresay it did mean something, that in my solitary walks for the preceding fourteen or fifteen years, a female companion often walked in fancy by my side, with whom I exchanged many a thought, and gave expression to many a feeling, and to whom I pointed out many a beauty in the landscape, and communicated many a curious fact, and whose understanding was as vigorous as her taste was faultless and her feelings exquisite.'

Mrs. Fraser finding out the state of affairs, and afraid that her daughter might form an alliance with a mechanic, interdicted the correspondence between the two for a time. The young lady was disconsolate at this. Her mother finding out that on the whole it would be more judicious to permit them to meet together, when things had gone thus far, removed the interdict, and they were again permitted to enjoy each other's society. An understanding was arrived at between them. They were to remain for three years more on the existing terms of intimacy, when, should no suitable field of exertion occur for Miller at home, they were then to quit the country for America. With a view of proving what he could do in the field of literature and editorial work, he resolved to publish his *Scenes and Legends of the North of Scotland; or, The Traditional History of Cromarty.* This was attended with some difficulty, but it was eventually published by Adam Black of Edinburgh. At this time occurred what he has termed one of the special providences of his life. The Commercial Bank of Scotland having decided to start a branch in Cromarty, Miller was offered the accountantship of the branch bank, without any security being required. This post he accepted, and travelled south to Edinburgh for instructions and initiation into the mysteries of banking. Linlithgow, a town west from Edinburgh, was chosen as the place where he would receive the training which was necessary for a bank agent. Looking backwards at this period, he says: 'I had wrought as an operative mason, including my term of apprenticeship, for fifteen years—no inconsiderable portion of the more active part of a man's life; but the time was not altogether lost. I enjoyed in these years fully the average amount of happiness, and learned to know more of the Scottish people than is generally known. Let me add—for it seems to be very much the fashion of the time to draw dolorous pictures of the condition of the labouring classes—that from the close of the first year in which I wrought as a journeyman up till I took final leave of the mallet and chisel, I never knew what it was to want a shilling; that my two uncles, my grandfather, and the mason with whom I served my apprenticeship—all working men—had had a similar experience; and that it was the experience of my father also.' He was also deeply conscious at this time of the change which had passed over him since meeting with Miss Fraser. He could write thus to her:—'How very inefficient, my L——, are the friendships of earth! My heart is bound up in you, and yet I can only wish and regret, and—yes, pray. Well, that is something. I cannot regulate your pulse, nor dissipate your pains, nor give elasticity to your spirits; but I can implore on your behalf the great Being who can. . . . I would fain be rich, that I might render you comfortable; powerful, that I might raise you to those high places of society which you are so fitted to adorn; celebrated, that the world might justify your choice.' After a five years' courtship, on the 7th January 1837 the two were happily united, when his salary, with a small addition from the earnings of his wife, who kept a few pupils, did not amount to much more than £100 a year.

He was now at Cromarty, a regular bank agent, and as usual investigating and looking about for any opening which might present itself, by which he could turn his leisure hours to account. Accordingly he contributed as many tales as would form a volume to the serial publication called Wilson's

Border Tales. This brought him about £25. The publica- of his *Scenes and Legends* had established his fame as a writer of vigorous prose. Leigh Hunt and Robert Chambers each spoke well of it in their respective journals; and Dr. Hetherington 'made it the subject of an elaborate and very friendly critique in the *Presbyterian Review*.' His life at this time he thus described in a letter to Mr. Robert Chambers :—' I am leading a quiet and very happy life in this remote corner, with perhaps a little less time than I know what to do with, but by no means over-toiled. A good wife is a mighty addition to a man's happiness ; and mine, whom I have been courting for about six years, and am still as much in love with as ever, is one of the best. My mornings I devote to composition ; my days and the early part of the evening I spend in the bank ; at night I have again an hour or two to myself ; my Saturday afternoons are given to pleasure—some sea excursion, for I have got a little boat of my own, or some jaunt of observation among the rocks and woods ; and Sunday as a day of rest closes the round.' Here he wrote several articles for *Chambers's Journal*. One or two of his sketches having been returned to him by the *Scottish Christian Herald*, he thus moralizes, in the style of Dr. Arnold, regarding the Society for the Diffusion of Useful Knowledge: 'I would fain see a few good periodicals set agoing of a wider scope than either those of the world or of the Church—works that would bear on a broad substratum of religion the objects of what I may venture to term a week-day interest. I can cite no book that better illustrates my beau-ideal of such a work than the Bible itself.' One of his own future coadjutors on the *Witness* newspaper, the late Dr. Andrew Cameron, helped to carry this wish into practice by originating the *Christian*

Treasury and the *Family Treasury;* but it remained for one of his own proteges, Mr. Alexander Strahan, to give even a broader and more distinct impulse to this type of literature in the different magazines *Good Words, Sunday Magazine,* and *Day of Rest*—all of which we owe to his genius and enterprise.

The question of spiritual independence was beginning to agitate the breasts of the people of Scotland. The patron of the parish of Auchterarder, Lord Kinnoul, had presented the Rev. Mr. Young to the charge. Only three individuals in the congregation had signed the call, three hundred had distinctly declined that he should be their pastor, and forty had remained neutral, and in this case the presbytery refused to sanction the ordination of Mr. Young. The Court of Session decided against this conclusion of the presbytery, and in May 1839 their decision was confirmed by the House of Lords. He wrote a pamphlet on the Non-intrusion controversy, which spread his fame, and became an introduction to the leaders of the movement in Edinburgh.

In 1839, Hugh Miller found himself in Edinburgh for the third time. The first time he had come as a journeyman mason in search of work, the second time to qualify himself for the bank agency, and the third time he arrived with his reputation made, and at the request of Mr. Robert Paul. He was introduced to Dr. Cunningham, Dr. Candlish, Dr. Abercromby, and others, and finally he was offered the editorship of a projected Non-intrusion newspaper, to be called the *Witness.* In spite of the confidence in his own powers which his recent success might have inspired, he was doubtful for a time of accepting it; but eventually he did so, and it was arranged that the *Witness* should start at the beginning of

1840. Of this turning-point in his career, he writes in his autobiography :—' I closed my connection with the bank at the termination of its financial year; gave a few weeks very sedulously to geology, during which I was fortunate enough to find specimens on which Agassiz has founded two of his fossil species; got, at parting, an elegant breakfast service of plate from a kind and numerous circle of friends, of all shades of politics and both sides of the Church; and was entertained at a public dinner, at which I attempted a speech, and got on but indifferently, though it looked quite well enough in my friend Mr. Carruthers' report, and which was, I suppose, in some sort apologized for by the fiddlers, who struck up at its close, "A man's a man for a' that." It was, I felt, not the least gratifying part of the entertainment that old Uncle Sandy was present, and that his health was cordially drunk by the company in the recognised character of my best and earliest friend. And then, taking leave of my mother and uncle, of my respected minister and my honoured superior in the bank, Mr. Ross, I set out for Edinburgh, and in a few days after was seated at the editorial desk—a point at which, for the present, the story of my education must terminate. I wrote for my paper during the first twelvemonth a series of geological chapters, which were fortunate enough to attract the notice of the geologists of the British Association, assembled that year at Glasgow, and which, in the collected form, compose my little work on the *Old Red Sandstone.* The paper itself rose rapidly in circulation, till it ultimately attained to its place among what are known as our first-class Scottish newspapers; and of its subscribers, perhaps a more considerable proportion of the whole are men who have received a university education than can be reckoned by any other Scotch journal of the same number of readers. And during the course of the first three years my employers doubled my salary. I am sensible, however, that these are but small achievements. In looking back upon my youth, I see, methinks, a wild fruit tree, rich in leaf and blossom; and it is mortifying enough to mark how few of the blossoms have set, and how diminutive and imperfectly formed the fruit is into which even the productive few have been developed. A right use of the opportunities of instruction afforded me in early youth would have made me a scholar ere my twenty-fifth year, and have saved to me at least ten of the best years of my life—years which were spent in obscure and humble occupations. But while my story must serve to show the evils which result from truant carelessness in boyhood, and that what was sport to the young lad may assume the form of serious misfortune to the man, it may also serve to show that much may be done by after diligence to retrieve an early error of this kind; that life itself is a school, and nature always a fresh study; and that the man who keeps his eyes and his mind open will always find fitting, though, it may be, hard schoolmasters, to speed him on in his lifelong education.'

These are noble words with which to close the record of his life up till this time. He lodged in St. Patrick Square, Edinburgh, until joined by Mrs. Miller and her daughter Harriet in April 1840, when they occupied a small house at No. 5 Sylvan Place, Meadows. Miller's salary at this time was £200; and as the sale of his household goods at Cromarty had only brought him £40, the furnishing of his house was only accomplished gradually. Mrs. Miller would herself occasionally contribute to the columns of the *Witness.* Mr. James Mackenzie, the sub-editor, was a great favourite with

Miller. Miller himself was never a ready leader-writer; but, as Dr. Chalmers remarked, 'when he did go off, he was a great gun, and the reverberation of his shot was long audible; but he required a deal of time to load.' His biographer, Dr. Peter Bayne, remarks regarding these leaders, that 'he meditated his articles as an author meditates his books, or a poet his verses, conceiving them as wholes, working fully out their trains of thought, enriching them with far-brought treasures of fact, and adorning them with finished and apposite illustration. . . . As complete journalistic essays, symmetrical in plan, finished in execution, and of sustained and splendid ability, the articles of Hugh Miller are unrivalled.' He conducted the newspaper, which was published twice a week, for sixteen years, and is said to have written no fewer than a thousand articles for its pages. There was a difference between him and some of the eminent Free Church leaders as to the style of conducting the *Witness*, which led to a private quarrel, in which Miller triumphed. It, however, made him shy of dealing with purely one-sided church affairs ever afterwards. It left him in proud isolation, and with little recognition from Free Church leaders, who were all the while reaping the benefit of his advocacy of their cause.

With reference to his powers of memory, Dr. Guthrie told the following story:—'We were sitting one day in Johnstone's (the publisher's) back shop, when the conversation turned on a discussion that had recently taken place in the Town Council on some matters connected with our church affairs. Miller said it reminded him of a discussion in Galt's novel of *The Provost*, and thereupon proceeded, at great length, to tell us what Provost this, and Bailie that, and Councillor the other said on the matter; but when he reached the

"Convener of the Trades," he came suddenly to a halt. Notwithstanding our satisfaction with what he had reported, he was annoyed at having forgotten the speech of the convener, and, getting a copy of the novel from the shelves in Johnstone's front shop, he turned up the place and read it, excusing himself for his failure of memory. But what was our astonishment, on getting hold of the book, to find that Miller had repeated pages almost verbatim, though it was some fifteen years or more since he had read the novel.'

More at home in the fields of literature and science, his *Old Red Sandstone*, which had appeared as a series of seven articles in the columns of the *Witness*, was issued in 1841. His charming book, *First Impressions of England and its People*, was the result of eight weeks' autumnal wandering in 1845, and the writing of it occupied six months of his editorial leisure. His fame was now established, and without ambition to shine in fashionable society, he politely declined all the invitations where he felt he would be out of his true sphere. A lady who met him shortly after his coming to Edinburgh as editor of the *Witness*, says:—' His appearance was that of a superior working man in his Sunday dress. His head was bent forward as he sat, but when he spoke he looked one full in the face with his sagacious and thoughtful eye. There was directness in all he said; to have spoken without having something to say would never have occurred to him. He had not the light, easy, inaccurate manner of speech one usually meets with : every word was deliberate, and might have been printed. There was a total want of self-assertion about him, but at the same time a dignified simplicity in the way he *placed his mind alongside* that of the person with whom he conversed. . . . His manner to women I always thought particularly good—wholly wanting in flattery, but full of gentle deference.

The greater portion of Hugh Miller's autobiography appeared in the *Witness* in 1853; it was published in the beginning of 1854, under the title of *My Schools and Schoolmasters*. In the spring of 1854, he lectured in Exeter Hall to the Young Men's Christian Association. His autumn holiday he always spent profitably, enlarging his knowledge of the geological features of Scotland. In the summer of 1855 he complained of weakness, and that his working power was not what it had been. He was troubled, too, with the lingering bad effects of the mason's disease. He began to carry pistols, as his imagination was haunted with the stories of robberies and outrages committed by desperate criminals, which were rife at that time. In the meantime he was labouring hard at the completion of his *Testimony of the Rocks*. Night after night, in spite of his wife's entreaties, he would return to his writing, and often only retired to rest in the early morning. Mrs. Miller, who was herself in poor health, aware that his nervous system was disordered, dreaded an attack of apoplexy. Two doctors were consulted, and it was found that he was suffering from an overworked mind, disordering his digestive organs, enervating his whole frame, and threatening serious head affection. His book was finished by this time. On the night of the 24th December 1856, he seems to have arisen from bed, and, in a paroxysm of madness, raised the thick woven seaman's jacket he wore over his chest, applied the muzzle of his revolver to his left side above the heart, and fired. The ball entered the left lung, grazed the heart, and cutting through one of the main arteries, lodged in the rib on the right side. The pistol slipping from his hand, fell into the bath close by. He had left in writing on a folio sheet of paper, which was lying on the table, the following words:—

'Dearest Lydia,—My brain burns. I *must* have walked; and a fearful dream rises upon me. I cannot bear the horrible thought. God and Father of the Lord Jesus Christ, have mercy upon me. Dearest Lydia, dear children, farewell. My brain burns as the recollection grows. My dear, dear wife, farewell. 'Hugh Miller.'

And 'so passed this strong heroic soul away.'

When Hugh Miller shot himself, Dr. Guthrie had been absent from home. 'On my return to the house next day,' he writes, 'I had two very painful duties to perform. The first was at the request of his eldest daughter, a very amiable as well as able young creature, to go up to the room where her father lay, and cut off a lock of his hair for her. I shall never forget the appearance of the body as I entered the room and stood alone by the dead: that powerful frame, built on the strongest model of humanity; that mighty head, with its heavy locks of auburn hair; and the expression of that well-known face, so perfectly calm and placid. The head was a little turned to one side, and the face thrown upwards, so that it had not the appearance of an ordinary corpse, but wore something of a triumphant, if not a defiant air, as if he were still ready for battle in the cause of truth and righteousness—defying his enemies to touch his great reputation as a man of the highest eminence in science, of the most unblemished character or the most extraordinary ability, and, more than any one of his compeers, entitled to be called a defender of the faith. The result of the *post-*

mortem examination showed that his reason had given way, and that he was in no way responsible for his acts.'

Letters of condolence to the bereaved family flowed in from all quarters, including heartfelt expressions of sympathy from Charles Dickens, Thomas Carlyle, and John Ruskin. His remains rest in the Grange Cemetery, Edinburgh, beside the dust of Dr. Chalmers; and twenty years afterwards, on the 11th March 1876, she who had been the true helpmate, the love and inspiration of his life, was laid beside him. His well-known works, in which there is so much of the pure gold of research and intellect, fill thirteen volumes. His wife and eldest daughter have also contributed to the field of light literature.

SIR TITUS SALT.

THERE have been those who have not scrupled to assert that large fortunes, and vast commercial interests, must of necessity have had some portion of falsehood or want of rectitude in their upbuilding; that an honest and at the same time a greatly successful business man, they are inclined to think, is rather an exceptional growth of this age and country. The man who claims our attention now, who had much to do in developing the commercial importance of the Midland Counties of England, and of the country itself, was both honest and God-fearing, and his life is another added to those perennial biographies whose lessons will be drawn upon by all right-thinking men for all time to come. The greatness of Sir Titus Salt was of a kind which would make him popular and useful anywhere; the poet, the author, and the preacher may appeal to a select few, but a man who appeals to the practical instincts of a practical people, as Sir Titus Salt has done, is sure to meet with almost universal understanding and approval.

The grandfather of Titus Salt bore the same name, and carried on an iron-founding business at Sheffield. He does not seem to have been particularly successful, as, when the business was handed over to his son Daniel Salt, he in turn carried it on but for a few years. This Daniel Salt was married on 5th July 1802, to Grace Smithies, of the Old Manor House, Morley. Her father had been a drysalter, and for a short period her young husband carried on this business. In personal appearance he has been described as a plain, blunt, Yorkshire man, with a strong muscular figure, and an impediment in his speech. In business he was noted for energy and industry. His wife was of a delicate constitution, of a retiring disposition, and with sweet manners. She belonged to a dissenting church, and was an earnest Christian woman.

Titus Salt, the first son of a large family, was born at the Old Manor House, Morley, Yorkshire, on the 20th September 1803. The village of his nativity, situated about four miles from Leeds, is said to have numbered but 2100 inhabitants at the time of his birth, although it has since grown until it numbers about 13,000. The people of the place had something of the old Puritan spirit about them, observing the Sabbath very strictly, keeping up the good habit of family worship, and were almost without exception Dissenters. Young Titus inherited his father's strong constitution, and in the words of one of his playmates, 'he was a bright boy for his years, full of fun when with those whom he knew well, but shy with strangers.' His first school was a dame school, where he learned to read; and in his ninth year, he attended a school at Batley in the neighbourhood. Batley was fully six miles from Morley, which was a very considerable walk for such a young lad. He carried his own dinner, consisting of oatcake and milk. This milk he was obliged to supply himself with before he left home, by milking the cow in the dark mornings.

Like many other great men, he was mostly indebted to his

mother for the higher elements of his home education; she instilled into his young mind a respect for religion, for the Sabbath, for the church, and for the Christian ministry, which remained with him through life. She taught him to pray and to read his Bible morning and evening. The following inscription was written on a Bible presented to him at this time. This same inscription he re-wrote on the Bibles which he presented to his own children :—

'TO TITUS SALT.

'May this blest volume ever lie
Close to thy heart and near thine eye;
Till life's last hour thy soul engage,
And be thy chosen heritage.'

In the year 1813, when in his tenth year, his father removed from Morley, and entered upon the work of a farm at Crofton, three miles from Wakefield. A young ladies' boarding-school kept there was presided over by Miss Mangall, the authoress of *Mangall's Questions*. At this time he attended a day school connected with Salem Chapel, Wakefield, kept by the Rev. B. Rayson. A letter from one of his schoolfellows contains the following sketch of his appearance at this time:—'Mr. Rayson gave up the school at Christmas 1815, from which time it was conducted by Mr. Enoch Harrison, who had for several years been Mr. Rayson's principal assistant, and with whom young Salt remained some time. His father's residence being upwards of three miles from school, Titus generally rode on a donkey, which was left until the afternoon at "The Nag's Head," a small inn near to the school, bringing with him in a little basket his dinner. In person he was tall and proportionately stout, and of somewhat heavy appearance. His dress was usually that of a country farmer's son,—viz. a cloth or fustian coat, corduroy breeches, with long gaiters, or, as they were generally called, "spats," or leggings, buttoned up the side, with strong boots laced in front. He was generally of a thoughtful, studious turn of mind, rarely mixing with his schoolfellows in their sports and play, and rather looked upon by them as the quiet, dull boy of the school. His words were generally so few that I cannot call to mind any particular thing that he either said or did. The school was a mixed school for both sexes, the boys occupying the ground floor and the girls the room above, and it was considered the best private day school in the town.' At this school he remained four years, and was well grounded in history, geography, and drawing. Mr. Harrison said of him, that 'he was never a bright pupil. He was very steady, very attentive, especially to any particular study into which he put his heart. Drawing was his chief delight. He was a fine, pure boy, stout and tall for his age, with a remarkably intelligent eye. So much did his eye impress me, that I have often, when alone, drawn it from memory, simply for my own gratification. I have sketches of him somewhere among my papers, with crimped frill round his neck, just as he appeared then; but though naturally very quiet, he was sometimes given to random tricks.'

His father did not succeed in the farm, but continued to lose money; when the lease expired, he removed to Bradford, when young Salt was in his nineteenth year. Bradford was just at that time entering on the career of remarkable prosperity for which it has since been so highly distinguished. The population of the town at that time was about 10,000; it has increased since to upwards of 170,000. His father, Daniel Salt, here began the business of a woolstapler, and in order that Titus might gain some knowledge of the same business, he was sent to the manufactory of Messrs. Rouse & Co., where he acquired a knowledge of wool-sorting and the other processes preparatory to weaving. Two brothers in the employment of this firm, named John and James Hammond, were of great service to him in teaching him the art of sorting wool.

His biographer, the Rev. R. Balgarnie, has thus described Titus Salt's daily work in the factory :—' He is a tall young man, with a "brat," or loose blouse, worn over his clothes to keep them clean; the fleece of wool is unrolled and spread out on the board. Being impregnated with natural grease, it holds entangled in its fibre a variety of substances with which the sheep while living had come into contact; these must be carefully removed. All the wool of the fleece is not of the same quality, but varies in length, fineness, and softness of fibre. It is the business of the sorter to separate these different qualities, and to put each into a basket. It is evident such occupation requires long and careful education, both of the eye and the hand. Had Titus Salt confined his attention exclusively to this one department of the business, and then at once joined his father, he might, perhaps, have been a successful woolstapler, but not a manufacturer; but, as we have said, he resolved to know every process, from the fleece to the fabric, and into each he put his heart. The next process was washing with alkali, or soap and water, and his knowledge of this served him in after years when his first experiments in alpaca began, and which he performed with his own hands. The next process was combing. It is necessary in the production of yarn that all the fibres should be drawn out and laid down smooth and distinct, and that all extraneous matters should be extracted. When Titus Salt was with the Rouses, this operation was done by hand; now, the combing machine, with its ingenious improvements, has superseded it, and become the glory of the trade. The wool thus combed is prepared for spinning.' When he had thoroughly mastered every detail of the business, he joined his father in the wool-stapling trade. The presence of a young, ardent, earnest spirit was soon felt in the business. His duty at first was the attending of the public wool sales in London and Liverpool, and in purchasing wool from farmers in Norfolk and Lincolnshire.

The church which the Salt family attended in Bradford was that of the Rev. Thomas Taylor, in Horton Lane. Titus Salt was enlisted in Sabbath-school work in connection with this church, and became first a librarian, then a teacher, afterwards a superintendent. From this time began the wholesome interest which he continued to take in Sunday schools throughout his lifetime; one of the last acts of his life was the erecting a school at Saltaire, at the cost of £10,000. In manner and appearance at this time, he has been described as 'very simple and quiet in his manner, not given to much speech, but a deep-thinking young man.'

A strike and disturbance taking place amongst the woolcombers of Bradford in 1825, Titus Salt used his influence in quelling the agitation, entering into the thick of the mob and endeavouring to bring them to reason. Although this attempt was in vain, it may be taken as a proof of his public spirit. At a later period, when a strike occurred amongst his own workpeople, and when waited upon by a deputation in order to settle the matter under dispute, he quietly replied, 'You are not in my service now. You have, of your own accord, left me; return to your work, and then I shall consider your proposals.' They did return to their work, and the matter was afterwards amicably settled.

The business conducted by Daniel Salt & Son continued to increase; they dealt mostly in worsted goods. To Titus Salt was due the credit of expanding the business beyond a mere local trade. As a commercial traveller, he possessed many good qualities. Calling at a warehouse in Dewsbury, a business man remarked, 'Mr. Titus Salt came to my warehouse one day, and wanted to sell wool. I was greatly pleased with the quiet power of the young man, and his aptitude for business, but most of all was I struck with the resolute way he expressed his intention of taking away with him, that day, £1000 out of Dewsbury.' Though styled the junior partner in the firm, by his practical knowledge he became in reality the head of the firm. His personal habits were becoming at this time confirmed: he was frugal and careful of his money, and avoided personal adornment. He made a vow with himself, which he kept, that he would not buy a gold watch until he had saved £1000. One most Christian and commendable rule he adopted at this time, was to devote a portion of his income to doing good through some religious channel.

When in Lincolnshire on business, and while calling on Mr. Whitlam at Manor House, Grimsby, he met his future wife. Caroline Whitlam, who afterwards became his wife, was the youngest of a large family of eighteen sons and daughters. 'You know,' he used to say when speaking of his love affairs, 'when I went courting, I made a mistake. It was another sister I was in quest of, but this one first met my eye, and captivated my heart at once.' The marriage took place in the Parish Church of Grimsby, on the 21st

August 1830. Titus Salt was in his twenty-seventh year, and the bride was but eighteen.

The practical and energetic mind of Titus Salt was not slow to profit from any new idea which would contribute to his success in business. He was the first to establish the fact that Donskoi wool, or wool from the banks of the river Don in Russia, could be used in the worsted as well as in woollen manufacture. He invested in this Russian wool, but could find no purchasers for it in its raw state. Accordingly, he took a mill, and fitting it up with suitable machinery, began to spin the Donskoi wool into yarn, and weave it into fabric. This experiment was entirely successful, and his business was now vastly increased. The like utilization of the wool called alpaca, and its service in the worsted trade, laid the basis of his fame and fortune. Being in Liverpool in the year 1836, the wool of the alpaca then first came under his notice. Passing through one of the dock warehouses, he saw a pile of dirty-looking bales of alpaca; the rents in the different bales disclosed their contents. Having examined a handful of the wool from one of the bales, he left the warehouse. Returning at a later time, he took away a small quantity of the material and brought it to Bradford. He had the alpaca wool scoured and combed, a process which he accomplished with his own hands, when he examined its fibre and measured its length. In this fibre he detected a thread which might be usefully utilized in the light fancy fabrics for which Bradford was noted. He mentioned the matter to his friend John Hammond. 'John, I have been to Liverpool and seen some alpaca wool; I think it might be brought into use.' But neither John Hammond nor his father encouraged him in his speculation. He remained firm, however, in his determination to give the wool a trial, and bought the whole consignment of alpaca from the Liverpool brokers at eightpence a pound. Next he organized new machinery for its manufacture into fabric.

Charles Dickens made this incident the basis of a humorous article in *Household Words*, part of which we quote :—

'A huge pile of dirty-looking sacks, filled with some fibrous material, which bore a strong resemblance to superannuated horse hair, or frowsy elongated wool, or anything unpleasant or unattractive, was landed in Liverpool. When these queer-looking bales had first arrived, or by what vessel brought, or for what purpose intended, the very oldest warehouseman in Liverpool docks couldn't say. There had once been a rumour—a mere warehouseman's rumour—that the bales had been shipped from South America on "spec," and consigned to the agency of C. W. & F. Foozle & Co. But even this seems to have been forgotten, and it was agreed upon all hands that the three hundred and odd sacks of nondescript hair wool were a perfect nuisance. The rats appeared to be the only parties who approved at all of the importation, and to them it was the finest investment for capital that had been known in Liverpool since their first ancestors had emigrated thither. Well, these bales seemed likely to rot or fall to the dust, or be bitten up for the particular use of family rats. Merchants would have nothing to say to them; dealers couldn't make them out; manufacturers shook their heads at the bare mention of them; while the agents, C. W. & F. Foozle & Co., looked at the bill of lading, and had once spoken to their head clerk about shipping them to South America again. One day—we won't care what day it was, or even what week or month it was, though things of far less consequence have been chronicled to the half minute — one day a plain, business-looking young man, with an intelligent face, quiet manner, was walking along through these same warehouses in Liverpool, when his eye fell upon some of the superannuated horse hair projecting from one of the ugly dirty bales. Some lady rat, more delicate than her neighbours, had found it rather coarser than usual, and had persuaded her lord and master to eject the portion from her resting-place. Our friend took it up, looked at it, felt at it, rubbed it, pulled it about; in fact, he did all but taste it, and he would have done that too if it had suited his purpose, for he was "Yorkshire." Having held it up to the light, and held it away from the light, and held it in all sort of positions, and done all sort of cruelties to it, as though it had been his most deadly enemy and he was feeling quite vindictive, he placed a handful or two in his pocket, and walked calmly away, evidently intending to put the stuff to some excruciating private torture at home. What particular experiments he tried with this fibrous substance I am not exactly in a position to state, nor does it much signify; but the sequel was, that the same quiet, business-looking young man was seen to enter the office of C. W. & F. Foozle & Co., and ask for the head of the firm. When he asked that portion of the house if he would accept eightpence per pound for the entire contents of the three hundred and odd frowsy dirty bags of nondescript wool, the authority interrogated felt so confounded that he could not have told if he were the head or tail of the firm. At first he fancied that our friend had come for the express purpose of quizzing him, and then that he was an escaped lunatic, and thought seriously of calling for the police, but eventually it ended in his making it over, in consideration of the price offered. It was quite an event in the little dark office of C. W. & F. Foozle & Co., which had its supply of light (of a very injurious quality) from the old, grim churchyard. All the establishment stole a peep at the buyer of the "South American stuff." The chief clerk had the curiosity to speak to him, and hear his reply. The cashier touched his coat-tails. The book-keeper, a thin man in spectacles, examined his hat and gloves. The porter openly grinned at him. When the quiet purchaser had departed, C. W. & F. Foozle & Co. shut themselves up, and gave all their clerks a holiday.'

The young manufacturer was now intensely busy, and he became closely devoted to his growing business. The demand for alpaca goods was increasing with great rapidity. Within three years the import had risen to 2,186,480 lbs., and since it has amounted to 4,000,000 lbs. The price had risen, too, from eightpence up to two shillings and sixpence. A great stimulus was given to trade in Bradford through this new industry. Thousands of workpeople received employment; they came from all parts of the country, and some foreigners, chiefly Germans, became resident workers of the community. Another successful combination at this time was the using a cotton thread in the worsted goods, which enabled the manufacturer to produce a lighter and cheaper article. For the next ten years after starting the manufacture of alpaca goods, he had a heavy burden upon him in superintending his different manufactories, which were situated in different parts of Bradford. His daily habit was to rise early, and he was generally in the warehouse before the engine was started. The people of Bradford had a saying to the effect that 'Titus Salt makes a thousand pounds before other people are out of bed.' His punctuality, too, was proverbial, and all his movements were regulated with the greatest accuracy. He also possessed in an eminent degree another admirable quality, which his biographer has called *whole-heartedness*. The work upon which he had now entered was carried forward with his whole heart.

The following sketch of the character of Titus Salt is by one of his workmen:—'He was a man of few words, but when he did speak, it was to the point, and pointed ; he meant what he said, and said what he meant. If I asked him for an advance of wages, he always said "I'll see," and it was done. He was a fair-dealing master between man and man. When he heard tell of a man trying to injure another man, that man had to go through the small sieve. If a man did his duty, he was always ready to give him a lift over the right. This I have myself proved. One day, Mr. Salt was coming down Manchester Road, Bradford, in his carriage. When he saw one of his workpeople, who had been ill for some time, he stopped his carriage and gave him a five-pound note. Whenever he saw true distress, he was always ready with his heart and hands to help them. He was a persevering, plodding man. He had a very strong struggle with the alpaca wool. It was, in some instances, thirty-six inches long ; but he was determined to master it, which he did.' It must be admitted, too, that he showed a considerable amount of public spirit beyond the sphere of his daily work. In the year 1832, he interested himself in railway communication with Leeds, in forming works to supply the town with water, and in the first Parliamentary election in the borough. As the last head constable of the town, Mr. Salt discharged his duties very thoroughly. In November 1848, he was elected Mayor of Bradford. The speech of Mr. Alderman Forbes, in proposing him for the office, gave in brief a good character sketch of the future Mayor.

'You are all, gentlemen, familiar,' he said, ' with Mr. Salt's character and position. The founder of his own fortune, he has raised himself to an eminence in the manufacturing interest of this town surpassed by none; and he now finds himself, as a reward for his industry, intelligence, and energy, at the head of a vast establishment, and affording employment to some thousands of workpeople. As we all know, Mr. Salt was the means of introducing a most important branch of trade into this town (I mean the alpaca trade), and thus rescuing that trade from comparative obscurity. Bringing to bear upon it his capital and skill, he not only realized great advantage for himself, but produced new fabrics in the manufactories of this district, thus developing a branch of business most important and beneficial to the working population. I believe, gentlemen, the same sagacity, practical good sense, cool judgment, and vigorous energy, which have hitherto distinguished Mr. Salt, will be brought to bear upon the public business of this borough. You need not be told of his princely benefactions to our various local charities, nor of that magnificent generosity which is always open to the appeal of distress, and the claims of public institutions having for their object the improvement of our population. With a warm heart, a sound head, a knowledge of our local interests conferred by long experience, and a disposition manifested on every occasion to do all that lies in his power to promote the prosperity of the borough, I do not think we could select a gentleman better qualified to succeed our late worthy Mayor, Robert Milligan, Esq.'

Mr. Salt was a warm admirer of Cobden and Bright, and at a banquet held to celebrate the abolition of the corn laws, he was called upon to acknowledge the toast of the corporation. That Bradford benefited and approved of these opinions, is evident from the fact that when the Exchange buildings were built, figures of Cobden, Salt, Gladstone, and Palmerston were placed round the outside. A white marble statue of Cobden has also been placed in the principal hall of the Exchange, which was unveiled by the Right Hon. John Bright.

The year 1848 was a period of great distress in Bradford. In one week, 17,680 lbs. of bread and 2954 quarts of soup were distributed over 1200 families. In January, previous to the French Revolution, he had been able to keep on most of his hands, but since that event his sales had fallen off £10,000. He was willing nevertheless to engage one hundred of the wool-combers who were unemployed, and lay their produce by.

Amongst the other benevolent movements in which Titus Salt was concerned, was that for establishing a Saturday half-holiday, and at his suggestion a meeting was convened in Bradford to think of some means to repress vice and profligacy. During a visitation of cholera, when several hundreds of deaths occurred, Mr. Salt contributed liberally to the wants of those who were suffering, and proved that he also sympathized with them by visiting many scenes of distress. When the tide turned and trade was again fairly prosperous, Titus Salt enabled 2000 of his workpeople to visit his summer residence at Craven, and breathe the wholesome air of the country. At the close of his year's mayoralty, the *Bradford Observer* wrote: 'Our worthy Mayor, Titus Salt, Esq., has long enjoyed wide-spread and well-merited popularity throughout this district. His kindness and consideration as an extensive employer, and his munificence and public spirit as an influential citizen, had long ago won for him " golden opinions from all sorts of men." He has lost none of his fame by the manner in which he has discharged the onerous duties of first magistrate of this borough, but has rather gained additional lustre to a good name.'

A mansion called Crow Nest, about seven miles west from Bradford, became the residence of Mr. Salt in 1844. He was obliged to drive to and from business, so that the time spent at home was very limited. His biographer takes great delight in recording that many a poor woman with a child in her arms, or many a dusty pedestrian, has had a lift from him by the way when driving. Two of his children, Whitlam and Mary, died here; their bodies were afterwards placed in the family mausoleum at Saltaire.

It had for a long time been a settled desire with him, that when he reached the age of fifty, he would dispose of his various mills, and spend the remainder of his life as a country gentleman. When the time came for an ultimate decision, it is said to have cost him many anxious days and sleepless hours by night. His habit of mind, which is certainly worthy of imitation, was to weigh a question calmly in his own mind, viewing it from all its different sides, and then to communicate his thoughts to others. The gigantic plan of bringing all his factories together on the banks of the Aire, it is said, was a gradual growth in his mind, and was first communicated to his friend Mr. Forbes. At last he decided on the site by the banks of the Aire. The following conversation is said to have taken place between the great manufacturer and his architect. On being shown the original draught of the plan, he examined it and shook his head.

Mr. Salt: 'This won't do at all.'

Mr. Lockwood: 'Pray, then, what are your objections to the sketch?'

Mr. Salt: 'Oh, it is not half large enough.'

Mr. Lockwood: 'If that is the only objection, I can easily get over it; but do you know, Mr. Salt, what this mill which I have sketched will cost?'

Mr. Salt: 'No; how much?'

Mr. Lockwood: 'It will cost £100,000.'

Mr. Salt: 'Oh, very likely.'

When other detailed plans were submitted to him, which with a few exceptions were adopted, his only remarks were the following:—

Mr. Salt: 'How much?'

Mr. Lockwood: 'About the sum I named before.'

Mr. Salt: 'Can't it be done for less?'

Mr. Lockwood: 'No, not in the way you want it to be done.'

Mr. Salt: 'Then let it be done as soon as possible.'

The erection of the buildings was proceeded with forthwith. They cover twenty-five acres, and the machinery is capable of turning out 30,000 yards of finished alpaca every day. Four thousand hands are employed in the works; and it is said he expended no less a sum than £100,000 on workmen's dwellings. The place contains 895 dwellings, without one single public-house. Mr. Salt erected a Congregational church in the centre of the buildings at a cost of £16,000.

'The Saltaire mills,' writes a good authority, 'are situated in one of the most beautiful parts of the romantic valley of the Aire. The site has been selected with uncommon judgment as regards its fitness for the economical working of a great manufacturing establishment. The estate is bounded by highways and railways, which penetrate to the very centre of the building, and is intersected by both canal and river. Admirable water is obtained for the use of the steam-engines, and for the different processes of the manufacture. By the distance of the mills from the smoky and cloudy atmosphere of a large town, unobstructed and good light is secured; whilst, both by land and water, direct communication is gained for the importation of coal and all other raw produce on the one hand, and for the exportation and delivery of manufactured goods on the other. Both porterage and cartage are entirely superseded; and every other circumstance which could tend to economize production has been carefully considered. The estate on which Saltaire is built will gradually develop itself to a considerable extent; and the part appropriated to the works, which is literally covered with buildings, is not less than six and a half acres in extent. Here the heavy operations of the manufacture are carried on; but the superficies given to the several processes and to the storage of goods, or, in other words, the floor area of the establishment, is in all about twelve acres. The main range of buildings, or the mill proper, runs from east to west, nearly parallel with the lines of railway from Shipley to Skipton and Lancaster. This pile is six storeys high, 550 feet in length, 50 feet in width, and about 72 feet in height; and the architectural figures, to avoid monotony, have been most skilfully treated by the architects. A bold Italian style has been adopted; and the beautiful quality of the stone of which the whole is massively built, displays its features to great advantage. Immediately behind the centre of the main mill, and at right angles to it, runs another six-storey building devoted to warehouse purposes, such as the reception and examination of the newly manufactured goods; and on either side of this, again, lie the combing shed (or apartment where the fibres of the alpaca, mohair, wool, etc., are combed by machinery), the handsome range of buildings devoted to offices, and the great shed for weaving by power-looms. It was in the combing shed that, in 1853, 3500 of Sir [then Mr.] Titus Salt's guests sat down to dinner, without confusion or crowding, and with perfect ventilation. The great loom shed would have accommodated under its single roof a party twice as numerous as this. Arranged in convenient situations are washing-rooms, packing-rooms, drying-rooms, and mechanics' shops. In the formation of the new roads which were requisite to secure free and easy access to the different parts of the mills, Sir Titus Salt availed himself of the most recent experience; therefore we find bridges of the most durable and solid construction, both in cast and wrought iron, one of these viaducts, on the tubular girder system, crossing the canal and river Aire, being not less than 450 feet in length.'

When Lord Harewood questioned him as to the change of his views in the building of Saltaire, he replied, 'My Lord, I had made up my mind to do this very thing, but on reflection I determined otherwise. In the first place,

I thought that by the concentration of my works in one locality, I might provide occupation for my sons. Moreover, as a landed proprietor, I felt I should be out of my element. You are a nobleman, with all the influence that rank and large estates can bring, consequently you have power and influence in the country; but outside of my business, I am nothing—in it, I have considerable influence. By the opening of Saltaire, I also hope to do good to my fellow-men.' The inauguration banquet in connection with the works came off on 20th September 1853, on Mr. Salt's fiftieth birthday, and also the date of the coming of age of his eldest son. The order given to the purveyors was for 3750 guests; two tons weight of meat, and half a ton of potatoes, were supplied for the occasion. The other provision on the occasion was most liberal and profuse. In his remarks on the occasion, Mr. Salt said substantially what we have already reported him as having said to Lord Harewood. A most successful concert was held in St. George's Hall, Bradford, in the evening.

Due regard was paid to the educational and religious wants of the community in Saltaire. The cost of the erection of the school buildings was £7000, and the Government Inspector's report upon them was: 'That the school buildings, for beauty, size, and equipment, had no rivals in the district.' Mr. Salt himself was closely attached to Congregationalism, but this did not prevent him taking an interest in and encouraging those of a different denomination. To the Wesleyans, and to the Primitive Methodists, he granted sites for their different places of worship. The Baptists have two chapels just outside the town, while the Roman Catholics and the Swedenborgians are also represented. An infirmary is also erected in the town for medical and surgical treatment. Any person maimed for life in the works receives a pension, or some light employment is given to him. Baths and wash-houses have also been erected at a cost of £7000. The wash-houses are provided with three steam-engines and six washing machines. Those who bring clothes to the place are provided with a rubbing and boiling tub, into which steam and hot and cold water are conveyed by pipes. The wringing machine is so contrived that the moisture is speedily expelled from the clothes, and the drying closet is heated with hot air. The clothes are then ready for the mangling and folding rooms. All the processes of washing can be gone through within an hour of the time the clothes are brought to the wash-house. The almshouses in the upper part of the Victoria Road are built in the style of Italian villas, and are capable of receiving seventy-five occupants. They were erected 'in grateful remembrance of God's undeserved goodness, and in the hope of promoting the comfort of some who, in feebleness and necessity, may need a home.' The occupants are understood to be unfit for labour, and may be either single or married men or women. A weekly allowance of ten shillings is given to a husband and wife, and seven shillings and sixpence to those who are unmarried. A neat little chapel has also been provided for them, where a service is held every Sunday, and also on Wednesday evening.

There was another great festival at the residence of Mr. Salt, at Crow Nest, on the 20th September 1856. This was his birthday, and also the anniversary of the opening of Saltaire Works. Three thousand of his workpeople visited Crow Nest, and on this occasion presented him in the evening with a colossal bust of himself, executed in Carrara marble, and on a pedestal of Sicilian marble. This large assemblage of workpeople visited, in the course of the day, the conservatories, the greenhouses, and enjoyed themselves with various sports in the park. No intoxicating liquors were provided, but the provisions for the occasion were of the amplest kind. The bill of fare was as follows:—Beef, 1380 lbs.; ham, 1300 lbs.; tongues and pies, 520 lbs.; plum bread, 1080 lbs.; currant bread, 600 lbs.; butter, 200 lbs.; tea, 50 lbs.; sugar, 700 lbs; cream, 42 gallons; and a great quantity of celery. The bust was presented to him in the evening in St. George's Hall, Bradford.

A Liberal in politics, Mr. Salt in the spring of 1869 came forward as a candidate for the borough of Bradford in the Liberal interest. He was elected, much to the gratification of his many friends, but only remained in Parliament two years. Writing to his constituents at the end of that time, he said:— 'I find, after two years of experience, that I have not sufficient stamina to bear up under the fatigues and late hours incident to Parliamentary life.' Though seldom absent from the House, he seldom spoke unless on some formal occasion. When down at Scarborough to recruit his health, he said to a friend, 'I am a weary man,' and he suffered not a little in health during his term of office.

Towards the end of 1856, Mr. Salt was obliged to remove from Crow Nest to Methley Park, after a residence there of seventeen years. He remained in the latter place for nine years. The place is six miles from Leeds, and before he could enter it, great alterations were necessary. Before leaving Crow Nest, he was presented with an Imperial Bible, elaborately bound, and with an address signed by the chief inhabitants of the district, expressive of their regret at his removal. The Rev. R. Balgarnie of Scarborough is inclined to date the turning-point in his inner life to the hearing of a sermon preached by himself, in his own church, from Isa. l. 4: 'The Lord God hath given me the tongue of the learned, that I should know how to speak a word in season to him that is weary.' Mr. Salt remarked to Mr. Balgarnie afterwards, 'That was a word in season to me yesterday; I am one of the weary in want of rest;' and after this time he showed a willingness to converse on spiritual subjects. The death of a favourite daughter, Fanny, came home to him with even greater power. Some time after this, he became earnestly decided in religious matters, and celebrated his first communion at Saltaire. 'This is the day,' he remarked in devout humility, 'I have long desired to see, when I should come and meet my people at the communion table.' From this time forward, he became still more truly and systematically benevolent than he had ever previously been. Liberal in the matter of denominational differences, he attended the Parish Church of Methley on Sabbath evenings very regularly. The bishop of the diocese applying to him to subscribe to some church-building scheme, he replied, 'I am a Nonconformist from conviction, and attached to the Congregational body. Nevertheless, I regard it as a duty and a privilege to co-operate with Christians of all evangelical denominations in furtherance of Christian work.' It is not known whether he subscribed in this instance or not, but it is well known that he subscribed to the fund for renovating York Minster, and he also presented the new Episcopalian church at Lightcliffe with a carved stone pulpit. Another act of munificence was the spontaneous gift of £5000 to the Sailor's Orphanage at Hull, and he continued to subscribe annually to its funds. Towards a Memorial Hall in connection with the Congregational Church, he also subscribed £5000. Towards the erection of South Cliff Congregational Church, Scarborough, he contributed in all £2500. He also gave a donation of £5000 towards the Lunatic Asylum for the Northern Counties, and £5000 to the Local Infirmary of Bradford; and to the treasurer of a Pastor's Retiring Fund he sent £1800.

His biographer draws a delightful sketch of his home life amongst his younger children at Methley. A number of distinguished guests were once gathered in his house, amongst whom were Owen Jones, Digby Wyatt, and Sir Charles Pasley. The conversation turning upon art and literature, the latter turned to their host, 'Mr. Salt, what books have you been reading lately?' 'Alpaca,' was his reply; and shortly afterwards he said, 'If you had four or five thousand people to provide for every day, you would not have much time for reading.' He had sufficient self-denial to give up smoking after it had become a confirmed habit. His chief delight at Methley was in the cultivation of fruits and flowers.

At Saltaire, public-houses were prohibited by the wise and generous founder, but this was more than compensated for by the erection of the Saltaire Club and Institute, at an expense of £25,000.

A dining hall for those who care to patronize it has also been erected opposite the works, where food can be obtained at a cheap rate. The other institutions and societies are thus summed up in his biography. These are—a fire brigade, a horticultural society with an annual show, a cricket club, a brass band, a string and reed band, a glee and madrigal society, an angling association, a co-operative and industrial society, a coal society, a funeral society, and other benefit societies for the sick.

In December 1867, Mr. Salt was enabled to return to his old residence at Crow Nest. We give a description of the place from his biography:—

CROW NEST.

'The mansion is of hewn stone, and consists of the centre portion, with a large wing on either side, connected by a suite of smaller buildings, in the form of a curve. . . . The conservatories are also situated on the south side, in a line with the mansion, and are so lofty and extensive as almost to dwarf its appearance. The central conservatory is more spacious than the others, and contains, in a recess, an elaborate rockery and cascade, of French workmanship, which were objects of great attraction at the Paris Exhibition. The lake was constructed after Mr. Salt's return, and affords another illustration of his fine eye for the beautiful and picturesque in nature. It is of uniform depth, well stocked with fish and aquatic birds, the latter finding shelter on the island in the middle. The vineries, pineries, and banana house are situated at a considerable distance from the mansion. We have previously stated that Mr. Salt took great delight in the cultivation of fruits and flowers, but the banana was his special favourite at the Crow Nest, and it attained dimensions rarely met with in this country. Its luxuriant foliage, immense height, and gigantic clusters of breadfruit more resemble those of a tropical than of a temperate clime.

'Let us enter the mansion itself. On the right hand of the entrance hall stands the colossal bust presented by the workpeople in 1856, close to which is the business-room, so called because it was used for the reception of visitors who called upon him for the transaction of business, or deputations for the presentation of appeals, etc. On the left hand is the morning-room, where he usually sat with his family, and from which a door opens into the spacious library, which is the largest and handsomest room of all. In the library is a beautiful bust of Mr. Salt, sculptured in white marble. This is the last delineation of his features, which have been well brought out by the artist, Mr. Adams-Acton. The dining, drawing, and billiard rooms are furnished with exquisite taste. And this is the scene to which Mr. Salt retired to spend the evening of his life.'

ONE DAY'S OCCUPATION AT CROW NEST.

'The hour of breakfast is eight o'clock, but before that time he has made his first appearance in the dining-room, where the lion's share of the post bag awaits him, containing, for the most part, applications from various parts of the country, and from all "sorts and conditions of men," for pecuniary aid. Perhaps one-half of them are appeals for building churches or schools, or for the liquidation of debts upon them; and the other half has a variety of wants to make known. One institution is restricted in its usefulness by want of funds, and much needs a helping hand; a widow is destitute, and the family cast upon the world; a young man wishes to go to college; a literary man is bringing out a book and wants it circulated; a deputation hopes to be allowed to present a " pressing case !" All these letters he briefly scans; but they are afterwards to be carefully perused and respectively answered. After breakfast the household assembles for morning prayer. The head of the house slowly reads a portion of sacred Scripture with much impressiveness, then prayer is solemnly read from the *Altar of the Household.* Thus the day is begun with God, and when evening comes it is closed in the same manner. Now the family separate to their respective duties. His occupation to-day is to answer the numerous letters that have arrived. In this important business his eldest daughter is his confidential secretary, which post she ably filled until the time of her marriage.'

These acts speak volumes, and show what a centre of influence he had become. In 1869 he received a note from Mr. Gladstone, then Prime Minister, that Her Majesty proposed that he should receive a baronetcy, which, after some consideration, he accepted. At the opening of the Congregational Church, Lightcliffe, in 1871, Dr. Guthrie, the Rev. Thomas Binney, and Newman Hall, LL.B., took part in the opening services, and were the guests of Sir Titus. It was on that occasion, when speaking at the public luncheon on the subject of ministers' stipends, that Dr. Guthrie expressed himself as follows :—'Some persons in Scotland,' he said, ' demur to this, because, in primitive times, ministers had not even a house, but wandered about in sheepskins and goatskins, being destitute, afflicted, tormented ! I asked them how they would like to see Candlish and me walking along the streets of Edinburgh in sheepskins and goatskins, horns and all.' The Saltaire public park is situated about five minutes' walk from the town, and was intended by Sir Titus to furnish wholesome recreation for the young and old. The area enclosed consists of 14 acres. The park was formally opened on the 25th July 1871. While Sir Titus was thus continually proving his thorough interest in his employees, tokens of gratitude and respect on their part, as we have seen, were not wanting. On 16th August 1871, he was presented with his portrait in the large hall of the Saltaire

Institute. His reply was, as usual, brief and pointed. 'I may now congratulate you and myself,' he said, 'on the completion of Saltaire. I have been twenty years at work, and now it is complete; and I hope it will be a satisfaction and a joy, and will minister to the happiness of all my people residing here. If I was eloquent, or able to make a long speech, I should try to do so, but my feelings would not allow me.' An event took place in the autumn of 1872 which moved him deeply; this was the marriage of his eldest daughter, who had for many years acted as his amanuensis and secretary, to Henry Wright, Esq., J.P., London.

On his seventieth birthday, and on the twentieth anniversary of the opening of Saltaire, he had the happiness of again seeing his workpeople assembled around him, to the number of 4200, at Crow Nest. In reply to a congratulatory speech he said: 'I am exceedingly glad to see all my workpeople here to-day. I like to see you about me, and to look upon your pleasant and cheerful faces. I hope you will all enjoy yourselves this day, and all get safely home again without accident after your day's pleasure. I hope to see you many times yet, if I am spared; and I wish health, happiness, and prosperity to you all.'

When it was proposed to erect a public statue to him in Bradford, the circular bearing on the business came under his notice, and, after reading it, he remarked: 'So they wish to make me into a pillar of Salt.' In the course of a personal interview which the committee had with him, he resolutely refused to sanction the movement, and ultimately all he would engage to do was to remain quiet and make no public announcement of his disapproval. Mr. John Adams-Acton, the sculptor, on receiving the order, proceeded to Carrara to secure a suitable block of marble. The block weighed fourteen tons, and required sixteen horses to convey it from the wharf. When finished, it showed him in a characteristic attitude, his right arm resting on the chair on which he was sitting, and holding a scroll in his left hand. The canopy above is from a design by Messrs. Lockwood & Mawson, and is in keeping with the architecture of the Town Hall in the immediate neighbourhood. The cost of the whole, which was covered by a general subscription, was £3000. The day of the unveiling of the statue was kept as a general holiday; the Duke of Devonshire conducted the ceremony, and spoke plainly and pointedly of the value and usefulness of the life he had lived. Amongst others, Mr. Morley, M.P., took part, and spoke as follows :—

SPEECH BY MR. MORLEY, M.P., AT THE UNVEILING OF THE STATUE.

'He was here to thank him for the stimulus of a noble example, and to express his thankfulness for this, that there is not a home in Great Britain that is not happier, more pure, with more comforts in it, owing to the continuous and earnest efforts made by enlightened and earnest men, amongst whom Sir Titus Salt had always held a prominent position. There had never been an object presented to him, that could tell in any way upon the wellbeing either of his neighbours or fellow-countrymen, which had not found in him a readiness to give either personal service or pecuniary help to the fullest extent required; and, therefore, he was entitled to the fullest expression of public gratitude, and their desire was, even while he is living, to show him that they were not unmindful of the services he had bestowed. In this money-loving and wealth-acquiring age, it was refreshing to find a man possessed of means, and glad of opportunities—almost thinking it a favour when opportunities were put before him—for dispensing the wealth which in so large a measure God had given him as the result of his own intelligent efforts. He might add that, as by conviction, and in obedience to conscience, Sir Titus Salt was a Nonconformist, he had never confined his princely liberality within the narrow limits of a mere sect, but had been ready, with a liberality of spirit which had always done him honour, to promote the erection of churches and schools, and the promotion of any organization whatever which, by God's blessing, might tell upon the material, social, and, above all, the religious wellbeing of the people among whom he has lived. There were thousands now before him, each one of whom might take a lesson from the life of this distinguished man. They might depend upon it that, when the History of England came to be written, a very substantial chapter would be given to the class of men of whom Sir Titus Salt was a distinguished ornament, and who, by personal sympathy and continuous earnest effort, have contributed so largely to the good work that has been done during the last forty years. There was need when such men were advancing in years, or passing away, for an accession of fresh men to come forward to carry on the work that had been so nobly begun. He commended, with all his heart, the example of Sir Titus Salt's life to the imitation of every inhabitant of the town.'

The last great public and benevolent act of Sir Titus Salt's life was the erection of a suite of buildings for the Sunday schools of Saltaire, one of the principal features of which is that the teaching halls consist of separate rooms. Every classroom is supplied with a small table and chair for the teacher's use, and the floors are covered with Brussels carpet.

A life-size portrait of Sir Titus Salt, the presentation of the teachers, hangs over the eastern gallery, within the building. The organ-harmonium within was presented by Mr. George Salt. The cost of the entire structure was £10,000, and the opening ceremony took place on the 30th May 1876.

His health steadily declined from the beginning of 1876. His walking exercise for the most part was confined to the library or garden terrace. He was still able, however, to pay several visits throughout the year amongst his friends. When unable to go to church or prayer meeting, his thoughts and prayers were with those who could. Being asked by his spiritual adviser on one occasion if his faith and hopes in Christ were clear and firm, he replied: 'No, not so much as I should like them to be; but all my trust is in Him. He is the only foundation on which I rest. Nothing else! Nothing else!' A visit paid to Scarborough in the autumn stimulated his waning strength a little at first, but he returned to Crow Nest a dying man. His life, as his biographer remarks, had been calm and orderly, and there remained little in his business or family interests which required setting in order. On Sunday, 17th December 1876, he was so ill that his absent children were telegraphed for. The end came on Friday afternoon, 29th December 1876. There was little physical suffering, and his spirit passed calmly away. Heartfelt letters of condolence flowed in from all quarters, and his remains were borne to their last resting-place in the family mausoleum amidst a concourse of 40,000 spectators.

His former pastor, the Rev. J. Thomson of Lightcliffe, in taking notice of the event on the following Sunday, preached a funeral sermon from Matt. xxv. 21. The following is an extract from the discourse:—

EXTRACT FROM FUNERAL SERMON BY REV. J. THOMSON.

'His greatness was the greatness of a great nature rather than of any separate or showy faculty. There was no meanness or littleness about anything he did. He lifted, by the sheer force of his own greatness, any matter in which he became vitally interested out of the realm of commonplace, and carried it irresistibly forward to final success. He moved without effort among great undertakings, liberal enterprises, and bountiful benefactions. What he did and gave was from the level in which he lived, and to which other men rise with effort, and only for a time. He could not be said to be a great reader, a great thinker, a great talker, a great expositor. He was better. He was a great man, having in him something responsive to all these forms of greatness; and standing among men, he was seen from afar; his very immobility, for it was the repose of strength, affording that support in trying times that gave a staying power to the undertakings with which he was identified, and which made them ultimately successful. Men knew always where to find him, and came also to trust that the cause to which he lent his name and influence had some just claims to consideration, and would finally succeed. In his personal friendships, where he trusted, he trusted wholly, and would not soon forsake one to whom he had given his confidence. The rising from one position to another in the social scale had no effect on his friendships. The friends of his youth were with him to the close; or, if not, it was they who had fallen asleep, or fallen away from him, and from these noble enterprises to which he had consecrated his strength and resources. He was a pioneer, a creator of the new era. He showed how the graces of the old feudalism that was being supplanted could be grafted on and exemplified by the men who brought forth and moulded the better age. No feudal lord could have set open his doors and offered his resources to the retainers of generations in the way he provided for those that laboured under his directions. The new era had, as it were, from the first a grace and benevolence that other social forms had never known, or known only in decay; and it owed, and owes, it to the personal characters of the men who laid its foundations, and not least to him whose removal we deplore. He was always seen to advantage among the people, surrounded by them, making his way among them, and through the path that they, with native courtesy, made for him. He treated them more at last as a benevolent and large-hearted father treats his children than as an employer treats his servants, or a leader his followers.'

There were other memorial services conducted in the name of the deceased, but his truest claim to remembrance rests with the manly rectitude of his life, with its Christian benevolence and unequalled manufacturing enterprise, which led to such legitimate success.

CHARLES DICKENS.

THE novelist plays an important part in modern life. There are hundreds of lives which must move on with little hope of change, living in the same street or village, and doing about the same round of duties every day. A work of fiction comes to those thus circumstanced as the revelation of a new world, expanding the hard lines of daily life, causing the mind to luxuriate in 'fresh fields and pastures new.' Besides, in a good novel, we get behind the scenes, and gain a glimpse of things as they really are, read the promptings to this or that act. As in real life, too, we meet with agreeable or disagreeable people, who may talk to us as they will, with this difference, that when we close the book we also for the time close the conversation, and lose sight of our company. To the tired man of business, the relaxation and 'play of mind' derived from a good novel are very welcome. There is much useless and damaging fiction, as there are many useless and wasted lives, but with a little selective power the bad may be avoided.

The novelist whose career we now follow, lived in his work and for his work, though with a keen zest for the other enjoyments of life. His writings have revealed much that was formerly hidden in what was formerly known as low life; and by their sunny humour and the development of odd and grotesque characters, he has created a body of literature by which the world has been amused and enriched. The faults and weaknesses of his books may be traceable to the faults and weaknesses in his character. Yet his life is one of deep and abiding interest to the student of biography.

The father of the popular novelist, John Dickens, was engaged as a clerk in the Navy Pay Office at Portsmouth. His wife's name was Elizabeth Barrow; she bore him eight children, two of whom died in infancy. Charles was the second youngest of the family, and was born at Landport, in Portsea, on Friday, 7th February 1812. Like Thomas de Quincey, reason dawned early, and his memory went back to the time when he was but two years of age. His father's duties caused the removal of the family to London in 1814, and shortly thereafter to Chatham. Young Dickens was at this time between four and five years of age. A house called Gadshill Place, between Rochester and Gravesend, took his attention and admiration, when his father told him that if he worked hard enough he might himself live in such a house. As a boy he was never strong; one good his early sicknesses did him, he believed, was the fact that they nourished an inclination for reading. To Washington Irving he afterwards spoke of himself as a 'very small and not-over-particularly-taken-care-of boy.' From his mother were received his first instructions in the art of reading. When the time came, along with his sister Fanny, he was sent to a preparatory day school. In this connection there is a passage in his novel, *David Copperfield*, which his biographer, Mr. John Forster, tells us is distinctly autobiographical.

Like many other children of good parts or rare genius, he began to write, and was precociously clever at singing comic songs; and at a very early age, also, he was taken to the theatre, which he enjoyed. The last two years of his residence at Chatham were spent at a school in Clover Lane, kept by a young Baptist minister, Mr. William Giles. His father's occupation led him to remove to London in 1821, and of the stage-coach journey thither he thus wrote:— 'There was no other inside passenger, and I consumed my sandwiches in solitude and dreariness, and it rained hard all the way, and I thought life sloppier than I expected to find it.' Shortly after their arrival in London, the Dickens family were involved in money difficulties which made retrenchment a necessity, when they resided in a poor locality, Bayham Street, Camden Town. His father was even arrested for debt and conveyed to Marshalsea prison, where the family followed. A walk through Covent Garden, the Strand, or Seven Dials, had a great attraction for him; one or two efforts at describing characters whom he met also belong to this period.

It may or it may not be creditable to the great novelist that there now occurred a passage in his life about which he never cared to speak, except to some most intimate friend; it never became public until his life was issued. This was the fact of his being sent, when about ten years of age, to make himself as useful as he could, under his cousin, George Lamert, in a blacking warehouse. His department was to cover the pots of paste-blacking with oil-paper, then with blue paper, to tie them round with a string, and then to trim them off neatly; next, when a sufficient quantity had been done, he affixed a printed label to them.

His great regret at this work consisted in the feeling that he was sinking in his companionships, and that his hopes of becoming learned and distinguished were doomed to disappointment. Proud he was, nevertheless, to be able to march home with six shillings he earned weekly in his pocket.

Some of the characters he met with while the Dickens family resided in Marshalsea prison were afterwards made use of in *Pickwick* and other novels. When they left this place, they removed to lodgings in Little College Street, and afterwards resided in Somers Town. He quitted the blacking warehouse when twelve years of age. Writing in 1862, he says: 'The never-to-be-forgotten misery of that old time bred a certain shrinking sensitiveness in a certain ill-clad, ill-fed child, that I have found come back in the never-to-be-forgotten misery of this later time.' But he never became a creature of circumstances; his early untoward surroundings only strengthened him to put forth the most determined and persevering energy to overcome them. His gift of animal spirit and sense of humour also helped to bear him up. In 1824 he went to a certain seminary called Wellington House Academy, kept by Mr. Jones, a Welshman, which he attended for two years. Mr. Thomas, one of his schoolfellows, says of this period: 'My recollection of Dickens whilst at school is that of a healthy-looking boy; small, but well built, with a more than usual flow of spirits, inducing to harmless fun, seldom, if ever, I think, to mischief, to which so many lads at that age are prone. I cannot recall anything that then indicated he would hereafter become a literary celebrity; but perhaps he was too young then. He usually held his head more erect than lads ordinarily do, and there was a general smartness about him.' Another schoolfellow says:

'He was a handsome, curly-headed lad, full of animation and animal spirits. . . . Depend on it, he was quite a self-made man, and his wonderful knowledge and command of the English language must have been acquired by long and patient study after leaving his last school.'

For a short time after leaving Wellington Academy, he attended another school; then he entered the office of Mr. Molloy in New Square, Lincoln's Inn, as a clerk, from which he removed to the office of Mr. Edward Blackmore, attorney, Gray's Inn. He entered this latter post in May 1827, and left in November 1828. His salary as an office lad at first amounted to thirteen shillings and sixpence, afterwards rising to fifteen shillings. The fact of his father having become a newspaper Parliamentary reporter for the *Morning Chronicle* may have decided him in the study of short-hand. 'The changes that were rung upon dots,' he writes, 'which in such a position meant such a thing, and in such another position something else entirely different; the wonderful vagaries that were played by circles; the unaccountable consequences that resulted from marks like flies' legs; the tremendous effects of a curve in a wrong place, not only troubled my waking hours, but reappeared before me in my sleep. When I had groped my way blindly through these difficulties, and had mastered the alphabet, there then appeared a procession of new horrors, called arbitrary characters — the most despotic characters I have ever known; who insisted, for instance, that the thing like the beginning of a cobweb meant expectation, and that a pen-and-ink sky-rocket stood for disadvantageous. When I had fixed these wretches in my mind, I found that they had driven everything else out of it; then, beginning again, I forgot them; while I was picking them up, I dropped the other fragments of the system; in short, it was almost heart-breaking.' All this painstaking labour became of real service in furthering his desire to get on, and at the age of nineteen he entered the reporters' gallery on behalf of the *True Sun*. Next he transferred his services to the *Mirror of Parliament*, and then to the *Morning Chronicle*.

His first published piece of writing saw the light in the *Old Monthly Magazine* for January 1834. This contribution, 'Mrs. Joseph Porter over the Way,' was stealthily dropped into a dark letter-box one night, and great was his exultation when it appeared in all the glory of print. He contributed in all nine sketches to this magazine, and began to use the signature of 'Boz.' This term was the nickname of his youngest brother, Augustus, whom, in honour of the *Vicar of Wakefield*, he had called Moses, but which, when pronounced through the nose, degenerated into Boz. 'To the wholesome training,' he afterwards said, 'of severe newspaper work, when I was a very young man, I constantly refer my first successes.' He wrote, in 1845, that 'there never was anybody connected with newspapers who, in the same space of time, had so much express and post-chaise experience as I. And what gentlemen they were to serve, in such things, at the old *Morning Chronicle!* Great or small, it did not matter. I have had to charge for the damage of a greatcoat from the drippings of a blazing wax candle, in writing through the smallest hours of the night in a swift-flying carriage and pair. I have had to charge for all sorts of breakages fifty times in a journey without question, such being the ordinary results of the pace which we went at. I have charged for broken hats, broken luggage, broken chaises, broken harness, everything but a broken head, which is the only thing they would have grumbled to pay for.' He has further said in allusion to this period: 'Returning home from exciting political meetings in the country to the waiting press in London, I do verily believe I have been upset in almost every description of vehicle known in this country. I have been, in my time, belated on miry by-roads, towards the small hours, forty or fifty miles from London, in a wheelless carriage, with exhausted horses and drunken postboys, and have got back in time for publication, to be received with never-forgotten compliments by the late Mr. Black, coming in the broadest of Scotch from the broadest of hearts I ever knew.' When the *New Monthly* ceased to pay Dickens for his sketches, he transferred this part of his services to the *Chronicle*, which remunerated him for them.

At the beginning of 1836, he found a publisher for the first series of *Sketches by Boz*, who offered him £150 for the copyright. In the *Times* of March 1836, the first announcement of the publication of the *Posthumous Papers of the Pickwick Club*, edited by Boz, was made. The then young publishing house of Messrs. Chapman & Hall had made overtures to him for this monthly serial, which was to contain certain sketches by Mr. Seymour, the artist, and which was also to contain an account of certain members of a Nimrod Club, who should go out shooting and fishing, and meet with sundry mishaps, owing to their inexperience and for other reasons. And thus the *Pickwick Papers* arose upon the world with their stores of amusement and laughter. Between the issue of the first and second numbers, Seymour, the artist, died by his own hand, but not before he had sketched the form and features of Mr. Pickwick as now so well known to the English public. His place in the illustration of the serial was supplied by Mr. Hablot K. Browne. His biographer, Mr. John Forster, who met him about this time, was charmed with his youthfulness and candid open countenance. He describes his forehead as good, a firm nose with full wide nostril, eyes beaming with intellect and cheerful humour. His whole face and bearing bore the stamp of quick, keen, practical power. Leigh Hunt said of it that it had 'the life and soul in it of fifty human beings.' His connection with the reporters' gallery was finished in 1836. In December of this year he wrote two pieces, the 'Strange Gentleman' and the 'Village Coquettes,' for the St. James' Theatre. On the 2d April of the same year, he had married Catherine, eldest daughter of Mr. George Hogarth, who had been a fellow-worker on the *Chronicle.*

The success of *Pickwick* was assured; it arose in parts from the comparatively small number of 400 to 40,000 copies. This success led Mr. Bentley, the publisher, to ask him to undertake the editorship of a new magazine which was to be started in January 1837. For this magazine he was to supply a serial story, and soon afterwards to write two other tales. Mr. Macrone, who had purchased the copyright of the first series of *Sketches by Boz*, now threatened to issue it in a serial form in the same way as *Pickwick*. This was only prevented by Messrs. Chapman & Hall buying up the copyright, with Dickens' consent, at the exorbitant price of £2000. The intimacy now begun with Mr. John Forster was of the greatest use to him; beyond a close friendship only closed with death, the latter read over the proof-sheets of his novels, making many important corrections, and often saving their author much trouble and anxiety when pressed with other work.

Dickens, to relieve the strain of hard work, began a habit at this time which continued throughout his lifetime, that of mental rest in bodily activity. He became a great walker, and pursued this steadily and systematically throughout a lifetime.

On his return from a holiday at Brighton in 1837, he was engaged in editing the life of Grimaldi the clown. For this book he wrote a preface, and re-told many of the stories, recasting them slightly from the materials which had been placed in his hands. The successful completion of *Pickwick* was the signal for a dinner, with himself in the chair, and T. N. Talfourd in the vice-chair. An agreement was entered upon with his publishers at this time, by which he was to succeed to a third ownership of the book; and at the same time another agreement was entered into for another work in parts, which was to run for nineteen months, for which he was to receive twenty several payments of £150, bringing the whole up to about £3000. For *Pickwick*, his respective payments must have exceeded £2500. The new novel just bargained for turned out to be *The Life and Adventures of Nicholas Nickleby*, the sale of the first number of which rose to 50,000. His previous arrangements made with Bentley hampered him not a little in his new work; however, *Oliver Twist* was finished by October 1838, and on publication it had a career of great popularity and success. It originally appeared in Bentley's *Miscellany*, for which he was still engaged to write another tale, *Barnaby Rudge*. He, however, managed to release himself from this engagement, handed over the editorship of the *Miscellany* to Mr. Harrison Ainsworth, and bought back the copyright and existing stock of *Oliver Twist* from Bentley for £2250.

In these days of growing prosperity, he took a cottage at Twickenham, where he spent the summer of 1838, enjoying the society of Talfourd, Thackeray, Jerrold, Edwin Landseer, George Cattermole, Stanfield, and W. H. Ainsworth. Before the close of 1839, he had removed from his old residence in Doughty Street into Devonshire Terrace, which was shut out from the new road by a high brick wall facing the York Gate into Regent's Park. His next novel was the *Old Curiosity Shop*, which was issued in weekly numbers. Of the first number, upwards of 70,000 copies were sold. He was to receive a payment of £50 for each number; the numbers were then to be accounted for separately, and half the realized profits paid to him, the other half going to the publishers. The writing of this story he felt intensely, and the death of little Nell affected him as powerfully as it could do an interested reader. *Barnaby Rudge*, begun during the progress of *Oliver Twist*, appeared in numbers in 1841. The prototype of Grip, the raven which plays such an important part in the story, was owned by himself, and he was much grieved at its death, which took place in 1841. In June 1841 he made a tour in Scotland, but took his work with him, transmitting regular instalments of the story he had in hand. On the 25th June, he was entertained at a public dinner in Edinburgh, one of the first and most striking acknowledgments that his genius was making itself known and at home amongst even the colder-blooded critical public in the northern capital. Professor Wilson, in the room of Lord Jeffrey, filled the chair, and made an admirable speech. Dickens, writing the day after the event to Mr. John Forster, said:—'The great event is over; and being gone, I am a man again. It was the most brilliant affair you can conceive; the completest success possible, from first to last. The room was crammed, and more than seventy applicants for tickets were of necessity refused yesterday. Wilson was ill, but plucked up like a lion, and spoke famously. I send you a paper herewith, but the report is dismal in the extreme. They say there will be a better one; I don't know where or when. Should there be, I will send it to you. I *think* (ahem!) that I spoke rather well. It was an excellent room, and both the subjects (Wilson and Scottish Literature, and the Memory of Wilkie) were good to go upon. There were nearly two hundred ladies present. The place is so contrived that the cross table is raised enormously, much above the heads of people sitting below; and the effect on first coming in (on me, I mean) was rather tremendous. I was quite self-possessed, however, and, notwithstanding the enthoosemoosy, which was very startling, as cool as a cucumber.' It may be interesting, and help to recall this scene, if we recall some of the speeches on this occasion.

Mr. Dickens said: 'If I felt your warm and generous welcome less, I should be better able to thank you. If I could have listened as you have listened to the glowing language of your chairman, and if I could have heard as you heard the "thoughts that breathe and words that burn" which he has uttered, it would have gone hard but I should have caught some portion of his enthusiasm, and kindled at his example. But every word which fell from his lips, and every demonstration of sympathy and approbation with which you received his eloquent expressions, render me unable to respond to his kindness, and leave me at last all heart and no lips, yearning to respond as I would do to your cordial greeting—possessing, Heaven knows, the will, and desiring only to find the way.

'The way to your good opinion, favour, and support has been to me very pleasing—a path strewn with flowers and cheered with sunshine. I feel as if I stood amongst old friends, whom I had intimately known and highly valued. I feel as if the deaths of the fictitious creatures in which you have been kind enough to express an interest, had endeared us to each other as real afflictions deepen friendships in actual life; I feel as if they had been real persons, whose fortunes we had pursued together in inseparable connection, and that I had never known them apart from you.

'It is a difficult thing for a man to speak of himself or of his works. But perhaps on this occasion I may, without impropriety, venture to say a word on the spirit in which mine were conceived. I felt an earnest and humble desire, and shall do till I die, to increase the stock of harmless cheerfulness. I felt that the world was not utterly to be despised; that it was worthy of living in for many reasons. I was anxious to find, as the Professor has said, if I could, in evil things, that soul of goodness which the Creator has put in them. I was anxious to show that virtue may be found in the by-ways of the world, that it is not incompatible with poverty and even with rags; and to keep steadily through life the motto expressed in the burning words of your northern poet:

> "The rank is but the guinea stamp,
> The man's the gowd for a' that."

And in following this track, where could I have better assurance that I was right, or where could I have stronger assurance to cheer me on, than in your kindness on this to me memorable night?

'I am anxious and glad to have an opportunity of saying a word in reference to one incident in which I am happy to know you were interested, and still more happy to know, though it may sound paradoxical, that you were disappointed —I mean the death of the little heroine. When I first conceived the idea of conducting that simple story to its termination, I determined rigidly to adhere to it, and never to forsake the end I had in view. Not untried in the school of affliction, in the death of those we love, I thought what a good thing it would be if in my little work of pleasant amusement I could substitute a garland of fresh flowers for the sculptured horrors which disgrace the tomb. If I have put into my book anything which can fill the young mind with better thoughts of death, or soften the grief of older hearts; if I have written one word which can afford pleasure or consolation to old or young in time of trial, I shall consider it as something achieved—something which I shall be glad to look back upon in after life. Therefore I kept to my purpose, notwithstanding that towards the conclusion of the story I daily received letters of remonstrance, especially from the ladies. God bless them for their tender mercies! The Professor was quite right when he said that I had not reached to an adequate delineation of their virtues; and I fear that I must go on blotting their characters in endeavouring to reach the ideal in my mind. These letters were, however, combined with others from the sterner sex, and some of them were not altogether free from personal invective. But, notwithstanding, I kept to my purpose, and I am happy to know that many of those who at first condemned me are now foremost in their approbation.

'If I have made a mistake in detaining you with this little incident, I do not regret having done so; for your kindness has given me such a confidence in you, that the fault is yours and not mine. I come once more to thank you, and here I am in a difficulty again. The distinction you have conferred upon me is one which I never hoped for, and of which I never dared to dream. That it is one which I shall never forget, and that while I live I shall be proud of its remembrance, you must well know. I believe I shall never hear the name of this capital of Scotland without a thrill of gratitude and pleasure. I shall love while I have life her people, her hills, and her houses, and even the very stones of her streets. And if in the future works which may lie before me you should discern—God grant you may!—a brighter spirit and a clearer wit, I pray you to refer it back to this night, and point to that as a Scottish passage for evermore. I thank you again and again, with the energy of a thousand thanks in each one, and I drink to you with a heart as full as my glass, and far easier emptied, I do assure you.'

Later in the evening, in proposing the health of Professor Wilson, Mr. Dickens said :—

'I have the honour to be entrusted with a toast, the very mention of which will recommend itself to you, I know, as one possessing no ordinary claims to your sympathy and approbation, and the proposing of which is as congenial to my wishes and feelings as its acceptance must be to yours. It is the health of our chairman, and coupled with his name I have to propose the Literature of Scotland—a literature which he has done much to render famous through the world, and of which he has been for many years, as I hope and believe he will be for many more, a most brilliant and distinguished ornament. Who can revert to the literature of the land of Scott and of Burns without having directly in his mind, as inseparable from the subject and foremost in the picture, that old man of might, with his lion heart and sceptred crutch, Christopher North? I am glad to remember the time when I believed him to be a real, actual, veritable old gentleman, that might be seen any day hobbling along the High Street, with the most brilliant eye—but that is no fiction—and the greyest hair in all the world; who wrote not because he cared to write, not because he cared for the wonder and admiration of his fellow-men, but who wrote because he could not help it, because there was always springing up in his mind a clear and sparkling stream of poetry which must have vent, and, like the glittering fountain in the fairy tale, draw what you might, was ever at the full, and never languished even by a single drop or bubble. I had so figured him in my mind, and, when I saw the Professor two days ago, striding along the Parliament House, I was disposed to take it as a personal offence: I was vexed to see him look so hearty. I drooped to see twenty Christophers in one. I began to think that Scottish life was all light and no shadows, and I began to doubt that beautiful book to which I have turned again and again, always to find new beauties and fresh sources of interest.'

In proposing the memory of the late Sir David Wilkie, Mr. Dickens said:—

'Less fortunate than the two gentlemen who have preceded me, it is confided to me to mention a name which cannot be pronounced without sorrow, a name in which Scotland had a great triumph, and which England delighted to honour. One of the gifted of the earth has passed away, as it were, yesterday; one who was devoted to his art, and his art was nature—I mean David Wilkie. He was one who made the cottage hearth a graceful thing—of whom it might truly be said that he found "books in the running brooks," and who has left in all he did some breathing of the air which stirs the heather. But however desirous to enlarge on his genius as an artist, I would rather speak of him now as a friend who has gone from amongst us. There is his deserted studio—the empty easel lying idly by—the unfinished picture with its face turned to the wall; and there is that bereaved sister, who loved him with an affection which death cannot quench. He has left a name in fame clear as the bright sky; he has filled our minds with memories pure as the blue waves which roll over him. Let us hope that she who more than all others mourns his loss, may learn to reflect that he died in the fulness of his fame, before age or sickness had dimmed his powers, and that she may yet associate with feelings as calm and pleasant as we do now the memory of Wilkie.'

Having arranged regarding the issue of another work of fiction, Dickens decided definitely regarding a visit to America, and sailed with his wife on the 3d of January 1842. After a stormy passage, he was everywhere received with the greatest enthusiasm.

On his return he issued *American Notes for General Circulation*, with a frontispiece by Clarkson Stanfield, R.A. Its publication gave great offence to his American readers. At the earnest request of his friend, Mr. John Forster, an introductory chapter was suppressed; this section was afterwards printed in his life, when all danger of its doing harm might be said to be over. Before the close of the year, four large editions had been sold. Mr. H. W. Longfellow, the poet, was his guest this year. A trip to Cornwall was also undertaken in the company of Mr. Forster, Stanfield, and Maclise. 'Such a trip we had into Cornwall,' he wrote to Mr. J. T. Fields, 'just after Longfellow went away. . . .

Sometimes we travelled all night, sometimes all day, sometimes both.... Heavens! If you could have seen the necks of bottles, distracting in their immense varieties of shape, peering out of the carriage pockets! If you could have witnessed the deep devotion of the post-boys, the wild attachment of the hostlers, the maniac glee of the waiters! If you could have followed us into the earthy old churches we visited, and into the strange caverns on the gloomy sea shore, and down into the depths of mines, and up to the tops of giddy heights, where the unspeakably green water was roaring I don't know how many hundred feet below! If you could have seen but one gleam of the bright fires by which we sat in the big rooms of ancient inns at night, until long after the small hours had come and gone!... I never laughed in my life as I did on this journey. It would have done you good to hear me. I was choking and gasping and bursting the buckle off the back of my stock all the way. And Stanfield got into such apoplectic entanglements that we were often obliged to beat him on the back with portmanteaus before we could recover him. Seriously, I do believe there never was such a trip. And they made such sketches, those two men, in the most romantic of our halting-places, that you would have sworn we had the spirit of beauty with us, as well as the spirit of fun.' By the 12th of November 1842, after a good deal of thinking and alteration, Dickens had decided on *Martin Chuzzlewit* as the title of his new novel, the first part of which appeared on January 1, 1843. The American portions of the book were considered violent exaggerations, but its chief intention was to call attention to the system of ship hospitals and to workhouse nurses. The book, when issued in a complete form, was dedicated to Miss Burdett Coutts. Sydney

Smith wrote to him during the progress of the work: 'Pecksniff and his daughters, and Pinch, are admirable—quite first-rate painting, such as no one but yourself can execute.' That he had openness of soul and liberality of mind sufficient to recognise and acknowledge merit in another, was made apparent by his pointing out to John Forster two stories in course of publication in *Blackwood's Magazine*, by George Eliot. 'Do read them,' he wrote; 'they are the best things I have seen since I began my course.' This year he presided at the opening of the Manchester Athenæum, when Mr. Cobden and Mr. Disraeli also assisted. He spoke on the education of the very poor. In the intervals of the composition of *Martin Chuzzlewit*, *A Christmas Carol* was written, which appeared in December, illustrated by John Leech. 'I can testify,' says John Forster, 'to the accuracy of his own account of what befell him in its composition, with what a strange mastery it seized him for itself; how he wept over it, and laughed, and wept again, and excited himself to an extraordinary degree; and how he walked thinking of it fifteen and twenty miles about the black streets of London, many and many a night, after all sober folks had gone to bed.'

The sale of *Chuzzlewit* in numbers, at its best, was never over 23,000, a great falling off from the *Curiosity Shop* and *Barnaby Rudge*, which sold over 70,000. Although before the close of the year he had received a sum of £726 for the sale of 15,000 copies of the *Carol*, yet he found himself in monetary difficulties. Determined in the meantime to break off his publishing relationships with Messrs. Chapman & Hall, his usual publishers, at the same time he concluded an agreement with Messrs. Bradbury & Evans, printers, which, upon an advance made to him of £2800, assigned them a fourth share in whatever he might write during the next eight years. Then he determined upon resting awhile, and concluded upon a holiday in Italy, with his family, in the following year.

In May 1844 he presided at the Annual Conversazione of the Polytechnic Institution, Birmingham, when he made an admirable speech. One of the really kind acts performed by Dickens this year was the writing of a preface to a volume issued by Newby, written by John Overs, a working man. The author was dying of consumption when it was published, and did not long enjoy any little profit or fame which it brought to him.

Before his departure to Italy, he was entertained at a dinner by his friends, at the 'Trafalgar,' Greenwich, on 19th June 1844, Lord Normanby being in the chair. Writing from Genoa on the 8th October, he betrayed a strong desire to get back again to the London streets, and gave an outline sketch of his forthcoming Christmas story, which he purposed to name *The Chimes*. 'I like more and more,' he said, 'my notion of making, in this little book, a great blow for the poor. Something powerful I think I can do, but I want to be tender too, and cheerful; as like the *Carol* in that respect as may be, and as unlike it as such a thing can be.'

The book was published at the close of the year, with illustrations by Maclise, Doyle, Leech, and Stanfield. It was not one of his greatest successes. As dramatized it became exceedingly popular. He complained that the writing of it had made his face white in a foreign land. 'My cheeks, which were beginning to fill out, have sunk again; my eyes have grown immensely large ; my hair is very lank; and the head inside the hair is hot and giddy.' At the close of June

1845, after doing Italy pretty thoroughly, he was again in London, with the idea for a new periodical edited by himself floating through his mind. His Christmas tale for 1845 was entitled the *Cricket on the Hearth*, and its sale at the first doubled that of his two preceding tales. In the autumn of the same year he appeared as an actor at St. James' Theatre in Ben Jonson's play, 'Every Man in his Humour.' His Captain Bobadil was so good that Leslie the artist took a portrait of him in that character. His biographer says that as manager, also, he was the life and soul of the affair. In turn he was stage-director, stage-carpenter, scene-arranger, property-man, prompter, and band-master. In October 1845 he was busy assisting in the arrangements for a new daily paper of Liberal politics. This paper eventually appeared under the title of the *Daily News*, while he was advertised as being at the head of the literary department. Dickens retired from this responsibility, however, after a few months' experience of it. The prospectus was written by himself, and it told how it would be kept free from party bias, and be devoted to the advocacy of all rational and honest means whereby wrong might be redressed, just right maintained, and the welfare of society be promoted. The letters which afterwards appeared under the title of 'Pictures from Italy,' he contributed to its columns. This latter book did not meet with great success. Dickens was succeeded in the editorship by his friend John Forster.

Two years had elapsed since the issue of *Martin Chuzzlewit*, when we find him busy planning a new book. Writing to the Countess of Blessington, he said: 'Vague thoughts of a new book are rife within me just now; and I go wandering about at night into the strangest places, according to my usual propensity at such a time, seeking rest and finding none.' In order to start fairly, he sought a home in Switzerland, establishing himself in a house at Lausanne, which he has thus described in writing to Douglas Jerrold : 'We are established here, in a perfect doll's house, which could be put bodily into the hall of our Italian palazzo; but it is the most lovely and delicious situation imaginable, and there is a spare bedroom, wherein we could make you as comfortable as need be. Bowers of roses for cigar smoking, arbours for cool punch-drinking, mountainous Tyrolean countries close at hand, piled-up Alps before the windows,' etc. The rent was to be £10 a month for half a year. 'The country,' he further wrote, 'is delightful in the extreme—as leafy, green, and shady as England; full of deep glens and branchy places (rather a Leigh Huntish expression), and bright with all sorts of flowers in profusion. It abounds in singing-birds, besides—very pleasant after Italy; and the moonlight on the lake is noble. Prodigious mountains rise up from its opposite shore (it is eight or nine miles across at this point), and the Simplon, the St. Gothard, Mont Blanc, and all the Alpine wonders are piled there in tremendous grandeur. The cultivation is uncommonly rich and profuse. There are all manner of walks, vineyards, green lanes, corn fields, and pastures full of hay. The general neatness is as remarkable as in England. There are no priests or monks in the streets, and the people appear to be industrious and thriving. French (and very intelligible and pleasant French) seems to be the universal language. I never saw so many booksellers' shops crammed within the same space as in the steep up-and-down streets of Lausanne.' Here he remained and worked for six months, beginning *Dombey & Son*, and writing also his Christmas tale, *A Battle of Life*, amid many difficulties and discouragements. '' The difficulty,' he wrote, 'of going at what I call a rapid pace is prodigious; it is almost an impossibility. I suppose this is partly the effect of two years' ease, and partly of the absence of streets and numbers of figures. I can't express how much I want these. It seems as if they supplied something to my brain which it cannot bear, when busy, to lose. For a week or a fortnight I can write prodigiously in a retired place (as at Broadstairs), and a day in London sets me up again and starts me. But the toil and labour of writing, day after day, without that magic lantern, is immense.'

The profits accruing to the author from his fresh publishing arrangements, with *Dombey and Son*, for the first half-year amounted to £2820. As the writing of *Dombey* occupied his whole attention, there was no Christmas book this year. On the 1st December 1847, he acted as chairman at a meeting of the Leeds Mechanics' Society, and on the 28th of the same month he opened the Glasgow Athenæum.

During this year it had been announced that Shakespeare's house at Stratford-upon-Avon was for sale. A public subscription was set afoot, and by means of readings by Macready and a grand performance at Covent Garden Theatre, the sum of £3000 was realized, sufficient to purchase it. In order to provide for the proper care and custody of the house, a course of amateur entertainments was given, Messrs. Charles Knight, Peter Cunningham, and J. P. Collier being directors of the general management, and Dickens being the stage-manager. The first performance took place at the Haymarket Theatre on 15th May 1848.

The summer of 1848 was passed in what his biographer terms strenuous idleness, while only the task of writing his Christmas book, *The Haunted Man*, lay ahead. Early in

1849, Dickens was busy with what proved to be the finest and most popular of his works, *David Copperfield.* Its sale in parts averaged 25,000 copies. While this book was making steady progress, and before the year was out, he had commenced another tale, *Bleak House.* This story was begun in his new residence, Tavistock House, in November 1851, and was finished at Boulogne in August 1853. Its average sale in parts was 30,000 copies, which rose to 40,000 copies. Writing on the 7th October from Broadstairs, he gave an outline of a proposal for a new periodical. 'My notion is a weekly journal, price either three halfpence or twopence—matter in part original, in part selected, and always having, if possible, a little good poetry. . . . Upon the selected matter, I have particular notions. One is, that it should always be a subject. For example, a history of piracy, in connection with which there is a vast deal of extraordinary, romantic, and almost unknown matter. A history of knight-errantry, and the wild old notion of the Sangreal. A history of savages, showing the singular respects in which all savages are like each other, and those in which civilised men, under circumstances of difficulty, soonest become like savages. A history of remarkable characters, good and bad, *in* history,—to assist the reader's judgment in his observation of men, and in his estimates of the truth of many characters in fiction.' This weekly miscellany made its appearance under the title of *Household Words,* on the 30th of March 1850. The first number contained the beginning of a story by Mrs. Gaskell. Amongst its original contributors have been John Forster, W. H. Wills (for upwards of twenty years its assistant editor), G. A. Sala, Moy Thomas, John Hollingshead, Miss Martineau, Professor Morley, Edmund Yates, Dr. Charles Mackay, and others. Dickens' editorial work was conscientiously done. The papers sent in, after some preliminary testing by the assistant editor, both MS. and proofs, received his careful attention.

During 1853 he felt conscious that he was overdoing it. Busy with *Bleak House*, the conduct of his new periodical, and the writing of his *Child's History of England*, he escaped from London to Boulogne on 13th June. 'If I had substituted,' he said, 'anybody's knowledge of myself for my own, and lingered in London, I never could have got through.' The completion of *Bleak House* was the signal for a trip to Italy, in company with Mr. Wilkie Collins and Mr. Augustus Egg. On his return to England, he began his career of public readings by giving his 'Christmas Carol' and 'The Cricket on the Hearth,' in the Birmingham Town Hall, in the middle of December. His success here strengthened his desire to become a public reader. Between four and five hundred pounds were added to the funds of the Institute through his exertions, and a prettily-worked flower basket in silver was at the time presented to Mrs. Dickens. The title of his next tale was *Hard Times*, which, out of a list of fourteen proposed titles, was the one to which he and Forster both agreed. It was the first tale which he contributed to his own magazine, *Household Words*. 'The difficulty,' he wrote, 'of the space, after a few weeks' trial, is crushing. Nobody can have any idea of it who has not had an experience of patient fiction-writing with some elbow-room always, and open places in perspective. In this form, with any kind of regard to the current number, there is absolutely no such thing.' John Ruskin characterized it as one of the most valuable of his novels. The name is said to have been originally derived from a tall, solitary brick house at Broadstairs; this watering-place for many years was Dickens' favourite seaside resort. The work itself in its purpose was directed against the Court of Chancery, for its enormous waste of time and costly procedure. For the advance sheets of the book, 2000 dollars are said to have been paid by Messrs. Harper Brothers, the New York publishers.

The work was inscribed, perhaps very justly, to Thomas Carlyle. This work was one of the least successful of his books with the general public. In a letter to Charles Knight acknowledging the receipt of a copy of his work *Knowledge is Power*, he described the aim of his book thus: 'My satire is against those who see figures and averages, and nothing else; the representatives of the wickedest and most enormous vice of this time; the men who, through long years to come, will do more to damage the really useful truths of political economy than I could do, if I tried, in my whole life; the addled heads who would take the average of cold in the Crimea during twelve months as a reason for clothing a soldier in nankeen on a night when he would be frozen to death in fur; and who would comfort the labourer, in travelling twelve miles a day to and from his work, by telling him that the average distance of one inhabited place from another on the whole area of England is not more than four miles. Bah! what have you to do with these?'

In October 1855, soon after the commencement of *Little Dorrit*, Dickens returned to London, to preside at a dinner given to W. M. Thackeray, previous to his departure for America on a lecturing tour. He made, as usual, a felicitous speech on the occasion. The first monthly portion of Thackeray's great novel, *Vanity Fair*, made its appearance on the 1st of February 1847, when the sunshine of critical and public favour dawned upon its author. But his writings, appealing to a more select public, never attained the same great circulation as those of Charles Dickens. Thackeray is reported to have said: 'Ah! they talk to me of popularity, with a sale of little more than one-half of 10,000. Why, look at that lucky fellow Dickens, with Heaven knows how many readers, and certainly not less than 30,000 buyers.' At another time he remarked to a friend, that it was very strange, yet nevertheless a fact, that Dickens' publishers sold five copies of his books for one which the booksellers sold of his own.

For his Christmas numbers 1854 and 1855, Dickens contributed the afterwards highly popular 'Richard Doubledick,' and 'Boots at the Holly Tree Inn.' In Christmas week 1855, Dickens read his 'Christmas Carol' to a large audience at the Mechanics' Institute, Sheffield, in aid of the funds. At the close, he was presented by the Mayor with a handsome table service of cutlery. The first number of *Little Dorrit* made its appearance at Christmas 1855, and the last in April 1857. It appeared at first in the usual twenty numbers, was issued when complete by Messrs. Bradbury & Evans, with illustrations by 'Phiz,' and a dedication to Clarkson Stanfield, R.A., the landscape painter. It showed up the procrastination and formal routine of the Government administration; its title originally stood as *Nobody's Fault*, the leading character who would bring about all the mischief in it laying the blame on Providence, and saying, 'Well, it's a mercy, however, nobody was to blame, you know.' Its sale in parts was over 35,000 copies, but it has not met with such continued popularity as some of his other works.

Early in 1856, Dickens made a change of residence from Tavistock House, Tavistock Square, to Gad's Hill Place. The account of how he came into possession of this place is thus pleasantly told in the *Daily News*:—'Though not born at Rochester, Mr. Dickens spent some portion of his boyhood there, and was wont to tell how his father, the late Mr. John Dickens, in the course of a country ramble, pointed out to him as a child the house at Gad's Hill Place, saying, " There, my boy; if you work and mind your book, you will, perhaps, one day live in a house like that." This speech sunk deep, and in after years, and in the course of his many long pedestrian rambles through the lanes and roads of the pleasant Kentish country, Mr. Dickens came to regard this Gad's Hill House lovingly, and to wish himself its possessor. This seemed an impossibility. The property was so held that there was no likelihood of its ever coming into the market; and so Gad's Hill came to be alluded to jocularly, as representing a fancy which was pleasant enough in dreamland, but would never be realized. Meanwhile the years rolled on, and Gad's Hill became almost forgotten. Then a further lapse of time, and Mr. Dickens felt a strong wish to settle in the country, and determined to let Tavistock House. About this time, and by the strangest coincidence, his intimate friend and close ally, Mr. W. H. Wills, chanced to sit next to a lady at a London dinner-party, who remarked, in the course of conversation, that a house and grounds had come into her possession of which she wanted to dispose. The reader will guess the rest. The house was in Kent, was not far from Rochester, had this and that distinguishing feature which made it like Gad's Hill and like no other place; and the upshot of Mr. Wills' dinner-table chit-chat with a lady whom he had never met before, was that Charles Dickens realized the dream of his youth, and became the possessor of Gad's Hill.'

Before leaving Tavistock House, Dickens gave a series of dramatic performances. His friend Wilkie Collins had written an entirely new drama for the occasion, called 'The Frozen Deep,' and a large room was fitted up with a stage, scenery, and footlights. Dickens' personation of one of the characters surprised all who witnessed it. By the death of Douglas Jerrold, in June 1857, Dickens lost an attached friend; he exerted himself on behalf of his widow, and in conjunction with Mark Lemon, Albert Smith, and others, a 'Jerrold Fund' was started. A series of entertainments was given, including a reading by Thackeray and Dickens at St. Martin's Hall, and a handsome sum was obtained. Lord Palmerston also granted to the widow from the Civil List an annual pension of £100 a year. Dickens' Christmas number for 1857 was founded on the Indian Mutiny, and was entitled, 'Perils of Certain English Prisoners.' An excursion into the Lake country with Wilkie Collins formed the basis of a series of articles, entitled, 'The Lazy Tour of Two Idle Apprentices.'

We come now to a passage in Dickens' domestic life which any writer might be glad to pass over lightly, but which plays such an important part in his life up to the close, and without which his true nature and character cannot be fully understood. From the letters sent to his biographer, Mr. Forster, from time to time, there were passages which betrayed great restlessness of nature. When remonstrated with for making a rush up Carrick Fell, in the Lake country, he wrote: 'Too late to say, "Put the curb on, and don't rush at hills:" the wrong man to say it to. I have now no relief but in action; I am become incapable of rest. I am quite confident I should rust, break, and die if I spared myself. Much better to die doing. What I am in that way, Nature made me first, and my way of life has of late, alas! confirmed. I must accept the drawback—since it is one—with the powers I have; and I must hold upon the tenure prescribed to me.' In writing regarding the plans for books floating through his mind, there is the same undertone of unrest. 'Am altogether in a dishevelled state of mind—motes of new books in the dirty air, miseries of older growth threatening to close upon me. Why is it that, as with poor David, a sense comes always crushing on me now, when I fall into low spirits, as of one happiness I have missed in life, and one friend and companion I have never made?' Then followed a complete account and disclosure of the skeleton in the domestic closet, which amounted to an apparently complete incompatibility between himself and his wife.

A friend presented Dickens with a Swiss châlet, which arrived from Paris in ninety-four pieces, fitting like a puzzle, and which formed a great resort to him during the summer months. In writing to an American friend, he said: 'I have put five mirrors in the châlet where I write, and they reflect and refract in all kinds of ways the leaves that are quivering at the windows, and the great fields of waving corn, and the sail-dotted river. My room is up among the branches of the trees, and the birds and the butterflies fly in and out, and the green branches shoot in at the open windows, and the lights and shadows of the clouds come and go with the rest of the company. The scent of the flowers, and indeed of everything that is growing for miles and miles, is most delicious.' The course of his life at Gad's Hill, unless when disturbed by visitors, was regular and methodical, as it had always been, his time being divided between working and walking. He enjoyed the dogs which from time to time he collected around him. Amongst his favourites was Turk, a noble mastiff, which, to its master's great grief, was killed in a railway accident. Linda, a real St. Bernard, brought by Mr. Albert Smith, grew to be a fine animal.

The great success of a reading which he gave at St. Martin's Hall, to assist the funds of a Sick Children's Hospital in Great Ormond Street, helped to determine the resolution which had been growing in his mind to give a series of public readings. The growing restlessness of his nature, too, arising from the circumstances previously mentioned, also helped this decision. A fortnight after this reading, he appeared in public in the character of a public reader on his own behalf. This was on the 29th April; and also about this time his old home was broken up, and he and his wife henceforward lived apart. The eldest son lived with his mother, the other children remained with himself, their intercourse with their mother being entirely left to themselves. A public statement regarding this altered relationship was made, as his biographer thinks, unwisely, in *Household Words*. His paid readings were given at the following dates: During 1858-59, in 1861-63, in 1866-67, and in 1868-70. The first series of readings was managed by Mr. Arthur Smith, the second by Mr. Headland, the third and fourth in America by Mr. George Dolby, acting for the Messrs. Chappell. The first series of readings, ending on the 27th of October 1859, comprised in all 125 readings. Beginning in St. Martin's Hall, they were continued during a provincial tour, embracing the chief English, Scotch, and Irish towns. Everywhere he was treated with the greatest personal affection and respect. At Liverpool, while on his way to Dublin, he had an audience of 2300 people. Besides the tickets sold, £200 in money was taken at the door. 'They turned away hundreds, sold all the books, rolled on the ground of my room knee-deep in checks, and made a perfect pantomime of the whole thing.' This reading, the 'Christmas Carol' and 'Pickwick,' had to be thrice repeated. In Dublin he was enthusiastically received, and was greatly pleased with the town and with its thriving, populous look. Of Belfast he remarked, ' A fine place, with a rough people; everything looking prosperous; the railway ride from Dublin quite amazing in the order, neatness, and cleanness of all you see; every cottage looking as if it had been whitewashed the day before; and many with charming gardens, prettily kept, with bright flowers. . . . . Enormous audiences. We turn away half the town. I think them a better audience on the whole than Dublin; and the personal affection is something overwhelming.' The net profit to himself for a time was £300 a week; in Scotland for one week it was much over this, being about £500. The subjects during his first reading tour were restricted to the 'Carol,' the 'Chimes,' the Trial in *Pickwick*, the chapters which contained 'Paul Dombey,' 'Boots at the Holly Tree Inn,' the ' Poor Traveller,' and ' Mrs. Gamp.'

Towards the end of 1857, he had presided at the fourth anniversary of the Warehousemen and Clerks' Schools; and in March 1858, in speaking at the Royal General Theatrical Fund Dinner at the Freemasons' Tavern, paid a graceful tribute to Thackeray, who presided on the occasion. In May he presided at the Artists' Benevolent Fund Dinner; and in July he took part at the opening of the Royal Dramatic College. In December he presided at the Institutional Association of Lancashire and Cheshire, in Manchester Free Trade Hall, distributing the prizes to candidates from 114 Mechanics' Institutes connected with the Association.

A dispute arose early in 1859 between Mr. Dickens and his publishers, which led to the discontinuance of *Household Words*, and also led to a return on his part to his old publishers. Messrs. Bradbury & Evans filed a bill in Chancery, and the winding up of the publication was directed, both parties refusing to sell their interest. Dickens owned five-eighths, and he had command over another eighth. The property on being put up to auction was bought back by Dickens for £3550.

What complicated the matter still further was the fact that Mr. Evans' son had married Miss Dickens. *All the Year Round* was the immediate successor of *Household Words*, and being an exact counterpart of the latter in all but the name, was immediately successful. Of the first quarter's statement regarding his new periodical, he wrote: 'So well has *All the Year Round* gone that it was yesterday able to repay me, with five per cent. interest, all the money I advanced for its establishment (paper, print, etc., all paid, down to the last number), and yet to leave a good £500 balance at the bankers.' The first number contained the opening of a new tale by himself, 'A Tale of Two Cities;' another of his novels, *Great Expectations*, was also contributed to its pages. Amongst the novelists who became contributors might be named Mr. Edmund Yates, Mr. Percy Fitzgerald, and Mr. Charles Lever. Mr. Wilkie Collins contributed his *Woman in White*, *No Name*, and *Moonstone*, and Charles Reade wrote for it his *Hard Cash*, and Lord Lytton his *Strange Story*. The sale of the extra Christmas numbers, before they were discontinued, was enormous, running as high as about 300,000 copies. A series of detached papers in the character of an *un*commercial traveller, contributed to this serial by the great novelist himself, had, many of them, a strong personal interest, and supplied many personal traits of both his earlier and later life to his biographer. He tells in one of them his cure for the disorder of sleeplessness, which was his 'turning out of bed at two, after a hard day, pedestrian and otherwise, and walking thirty miles into the country to breakfast.' For a short story contributed to the *New York Ledger*, called 'Hunted Down,' he received the large sum of one thousand pounds; a 'Holiday Romance,' and 'George Silverman's Explanation,' of the same length, were written for an American child's magazine issued by Mr. Fields, and for the same price.

The success of Dickens' second series of readings was as great and well assured as the first. Writing from Glasgow on the 3d December 1861, he described the following strange scene :—'Such a pouring of hundreds into a place already full to the throat, such indescribable confusion, such a rending and tearing of dresses, and yet such a scene of good humour on the whole, I never saw the faintest approach to. While I addressed the crowd in the room, G. addressed the crowd in the street. Fifty frantic men got up in all parts of the hall and addressed me all at once. Other frantic men made speeches to the walls. The whole B. family were borne in on the top of a wave, and landed with their faces against the front of the platform. I read with the platform crammed with people. I got them to lie down upon it, and it was like some impossible tableau or gigantic picnic—one pretty girl in full dress lying on her side all night, holding on to one of the legs of my table! It was the most extraordinary sight. And yet, from the moment I began to the moment of my leaving off, they never missed a point, and they ended with a burst of cheers.' A tempting offer, which could not be accepted, was made to him at this time by a gentleman in London, who offered him £10,000 for eight months' readings in Australia.

The Christmas number which Dickens issued for 1859 was called *The Haunted House*, and consisted of seven ghost stories. In August 1860, the two portions of his story, *Hunted Down*, appeared. The Christmas number for 1860 was called *A Message from the Sea*. The Christmas number for 1861 was called *Tom Tiddler's Ground*.

The sudden death of W. M. Thackeray on the Christmas eve of 1863, drew forth a graceful tribute from Dickens, which appeared in the *Cornhill Magazine* for February 1864.

The Christmas number of 1862 was entitled *Somebody's Luggage*, and was devoted to the interests of waiters. The number for Christmas 1863, which proved exceedingly popular, was entitled *Mrs. Lirriper's Lodgings*. His mother, who had been in infirm health for years, died in September 1863, and his own son Walter died on the last day of the same year, in the officers' hospital at Calcutta. He was a lieutenant in the 26th Native Infantry Regiment, and had been previously doing duty with the 42d Highlanders. His new story in twenty numbers, *Our Mutual Friend*, was now commenced. Number I. was published on the 1st of May 1864, with illustrations by Mr. Marcus Stone. A severe attack of illness in February 1865 left behind it a lameness in his left foot, which never afterwards wholly left him. This was attended with great suffering, while all the time he still persisted in his ordinary exercise in all weathers. During the summer he took a brief holiday in France. 'Before I went away,' he wrote, 'I had certainly worked myself into a damaged state. But the moment I got away, I began, thank God, to get well. I hope to profit by this experience, and to make future dashes from my desk before I want them.' His constitution, too, received a severe shock from a railway accident which happened at Staplehurst. The carriage in which he was journeying was thrown off the rails, and for a time hung suspended in the air. He had just time to scramble out of the window unhurt. Although feeling unwell, at the end of February 1866 he closed with an offer for a third series of readings for Messrs. Chappell, Bond Street, of £50 a night for thirty nights. He engaged to read in England, Ireland, Scotland, or Paris, and the fatigue which he afterwards underwent in journeying about so rapidly from place to place was immense. The sum taken amounted to £4720. His success was beyond even his former successes. Writing from Liverpool about the close of April, he remarked the sudden death of Mrs. Carlyle, which had taken place on the 2d of the same month: 'It was a terrible shock to me, and poor dear Carlyle has been in my mind ever since. How often I have thought of the unfinished novel. No one now to finish it. None of the waiting women come near her at all.' The novel referred to was a story in which the deceased had been engaged. A fresh negotiation was entered into with Messrs. Chappell in August 1866 for another course of readings; forty-two nights for £2500.

An instance of the cordiality of feeling subsisting between Dickens and the staff of his weekly periodical is furnished in his interesting story regarding the contributions of Miss Adelaide Anne Procter, the daughter of 'Barry Cornwall.' He wrote a touching preface for her *Legends and Lyrics*, which was issued after her death, and explained how he first gained her acquaintance.

'In the spring of the year 1853,' he writes, 'I observed, as conductor of the weekly journal *Household Words*, a short poem among the proffered contributions, very different, as I thought, from the shoal of verses perpetually passing through the office of such a periodical, and possessing much more merit. Its authoress was quite unknown to me. She was one Miss Mary Berwick, whom I had never heard of; and she was to be addressed by letter, if addressed at all, at a circulating library in the western district of London. Through this channel Miss Berwick was informed that her poem was accepted, and was invited to send another. She complied, and became a regular and frequent contributor. Many letters passed between the journal and Miss Berwick, but Miss Berwick herself was never seen. How we came gradually to establish, at the office of *Household Words*, that we know all about Miss Berwick, I have never discovered. But we settled, somehow, to our complete satisfaction, that she was governess in a family; that she went to Italy in that capacity, and returned; and that she had long been in the same family. We really knew nothing whatever of her, except that she was remarkably business-like, punctual, self-reliant, and reliable; so I suppose we insensibly invented the rest. For myself, my mother was not a more real personage to me than Miss Berwick the governess became. This went on until December 1854, when the Christmas number, entitled *The Seven Poor Travellers*, was sent to press. Happening to be going to dine that day with an old and dear friend, distinguished in literature as Barry Cornwall, I took with me an early proof of that number, and remarked, as I laid it on the drawing-room table, that it contained a very pretty poem, written by a certain Miss Berwick. Next day brought me the disclosure that I had so spoken of the poem to the mother of its writer, in its writer's presence; that I had no such correspondent in existence as Miss Berwick; that the name had been assumed by Barry Cornwall's eldest daughter, Miss Adelaide Anne Procter.'

He thus describes the final ending. She had then lain an invalid upon her bed through fifteen months:—'In all that time, her old cheerfulness never quitted her. In all that time, not an impatient or querulous minute can be remembered. At length, at midnight on the 2d of February 1864, she turned down a leaf of a little book she was reading, and shut it up. The ministering hand that had copied the verses into the tiny album was soon around her neck, and she quietly asked, as the clock was on the stroke of one, "Do you think I am dying, mamma?"—"I think you are very, very ill to-night, my dear." "Send for my sister. My feet are so cold. Lift me up!" Her sister entering as they raised her, she said, "It has come at last!" and with a bright and happy smile looked upward and departed.'

The publication of Dickens' last complete novel began in May 1864, and extended to November 1865. It was issued in the old twenty-number form, a shape which might on the whole yield a larger pecuniary return than the ordinary orthodox three-volume novel. Of *Our Mutual Friend*, John Forster, his biographer, remarks:—'When somewhat tired in September 1865 from the labour of writing this tale, he turned to his new Christmas number, and produced the delightful *Doctor Marigold's Prescriptions.* He wrote: "Tired with *Our Mutual*, I sat down to cast about for an idea, with a depressing notion that I was, for the moment, overworked. Suddenly the little character that you will see, and all belonging to it, came flashing up in the most cheerful manner, and I had only to look on and leisurely describe it." Before his last visit to America, he wrote three other Christmas pieces, "Barbox Brothers," "The Boy at Mugby Station," and "No Thoroughfare." The latter piece was written conjointly with Mr. Wilkie Collins.'

The last visit made by Dickens to America, from November 1867 to April 1868, was one long triumph. A farewell banquet was held on 2d November, in the Freemasons' Tavern, London; the company numbered between four and five hundred gentlemen.

Lord Lytton presided, and in the course of an eulogium upon the illustrious novelist said:—'We are about to entrust our honoured countryman to the hospitality of those kindred shores in which his writings are as much household words as they are in the homes of England.

'If I may speak as a politician, I should say that no time for his visit could be more happily chosen. For our American kinsfolk have conceived, rightly or wrongly, that they have some recent cause of complaint against ourselves; and out of all England, we could not have selected an envoy—speaking not on behalf of our Government, but of our people—more calculated to allay irritation and propitiate goodwill.

. . . .

'How many hours in which pain and sickness have changed into cheerfulness and mirth beneath the wand of that enchanter! How many a hardy combatant, beaten down in the battle of life—and nowhere on this earth is the battle of life sharper than in the commonwealth of America—has taken new hope, and new courage, and new force from the manly lessons of that unobtrusive teacher.'

He concluded by proposing 'A prosperous voyage, health, and long life to our illustrious guest and countryman, Charles

Dickens.' The reports given of the banquet described how the company rose as one man to do honour to the toast, and drank it with such expressions of enthusiasm and goodwill as are rarely to be seen in any public assembly. Again and again the cheers burst forth, and it was some minutes before silence was restored.

Mr. Dickens replied in a speech such as no one else could have delivered, and towards its conclusion he said:—'The story of my going to America is very easily and briefly told. Since I was there before, a vast and entirely new generation has arisen in the United States. Since that time, too, most of the best known of my books have been written and published. The new generation and the books have come together, and have kept together, until at length numbers of those who have so widely and constantly read me, naturally desiring a little variety in the relations between us, have expressed a strong wish that I should read myself. This wish, at first conveyed to me through public as well as through business channels, has gradually become enforced by an immense accumulation of letters from private individuals and associations of individuals, all expressing in the same hearty, homely, cordial, unaffected way a kind of personal affection for me, which I am sure you will agree with me that it would be downright insensibility on my part not to prize. Little by little this pressure has become so great that, although, as Charles Lamb says, " My household gods strike a terribly deep root," I have driven them from their places, and this day week, at this hour, shall be upon the sea. You will readily conceive that I am inspired besides by a natural desire to see for myself the astonishing progress of a quarter of a century over there; to grasp the hands of many faithful friends whom I left there; to see the faces of a multitude of new friends upon whom I have never looked; and though last, not least, to use my best endeavours to lay down a third cable of intercommunication and alliance between the Old World and the New.

"Twelve years ago, when, Heaven knows, I little thought I should ever be bound upon the voyage which now lies before me, I wrote, in that form of my writings which obtains by far the most extensive circulation, these words about the American nation: "I know full well that, whatever little motes my beamy eyes may have described in theirs, they are a kind, large-hearted, generous, and great people." In that faith I am going to see them again. In that faith I shall, please God, return from them in the spring, in that same faith to live and to die. My lords, ladies, and gentlemen, I told you in the beginning that I could not thank you enough, and, Heaven knows, I have most thoroughly kept my word. If I may quote one other short sentence from myself, let it imply all that I have left unsaid, and yet deeply feel; let it, putting a girdle round the earth, comprehend both sides of the Atlantic at once in this moment. As Tiny Tim observed, "God bless us every one."'

He arrived in Boston, safe and well, on the night of Tuesday, 19th November. His first reading came off at Boston on the 2d December, and was thus described by himself:—'It is really impossible to exaggerate the magnificence of the reception or the effect of the reading. The whole city will talk of nothing else and hear of nothing else to-day. Every ticket for those announced here and in New York is sold. All are sold at the highest price, for which in our calculation we made no allowance; and it is impossible to keep out speculators who immediately sell at a premium. At the decreased rate of money, even, we had above £450 English in the house last night; and the New York hall holds five hundred people more. Everything looks brilliant beyond the most sanguine hopes, and I was quite as cool last night as though I were reading at Chatham.' After a few readings at Boston, he left for New York, Washington, and Philadelphia, reading to immense audiences, and being received with great enthusiasm everywhere. On the 11th December, he wrote to his daughter from New York, 'Amazing success! A very fine audience, far better than at Boston. "Carol" and "Trial" on first night, great; still greater "Copperfield" and "Bob Sawyer" on second. For the tickets of the four readings of next week there were, at nine o'clock this morning, three thousand people in waiting, and they had begun to assemble in the bitter cold as early as two o'clock in the morning.' He had some acute and pertinent remarks to make regarding the New York newspapers! 'The *Tribune* is an excellent paper; Horace Greeley is editor-in-chief, and a considerable shareholder too. All the people connected with it whom I have seen are of the best class. It is also a very fine property; but here the *New York Herald* beats it hollow, hollow, hollow! Another able and well-edited paper is the *New York Times*. A most respectable journal, too, is Bryant's *Evening Post*, excellently written. There is generally a much more responsible and respectable tone than prevailed formerly, however small may be the literary merit, among papers pointed out to me as of large circulation. In much of the writing there is certainly improvement, but it might be more widely spread.' The reading of 'Doctor Marigold,' in New York, in January

1868, was a great hit. Dickens described his manager as always 'going about with an immense bundle that looks like a sofa cushion, but is in reality paper money, and it had risen to the proportions of a sofa on the morning he left for Philadelphia. Well, the work is hard, the climate is hard, the life is hard; but so far the gain is enormous. My cold steadily refuses to stir an inch. It distresses me greatly at times, though it is always good enough to leave me for the needful two hours. I have tried allopathy, homœopathy, cold things, warm things, sweet things, bitter things, stimulants, narcotics, all with the same result. Nothing will touch it.' This cold persistently remained with him during this fatiguing and exciting time, and he suffered greatly from sleeplessness. He usually breakfasted upon an egg and a cup of tea, had a small dinner at three o'clock, a small quail or something equally light when he came home at night. An egg beaten up in sherry before he began to read, and the same between the parts, was his other refreshment.

Before leaving New York, there were five farewell nights; 3298 dollars were the last receipts. A public dinner was given in his honour at New York, Horace Greeley occupying the chair. Dickens attended only with difficulty, and spoke with pain.

'It has been said in your newspapers,' he remarks, 'that for months past I have been collecting materials for and hammering away at a new book on America. This has much astonished me, seeing that all that time it has been perfectly well known to my publishers, on both sides of the Atlantic, that I positively declared that no consideration on earth should induce me to write one. But what I have intended, what I have resolved upon (and this is the confidence I seek to place in you), is, on my return to England, in my own person to bear, for the behoof of my countrymen, such testimony to the gigantic changes in this country as I have hinted at to-night. Also to record that, wherever I have been, in the smallest places equally with the largest, I have been received with unsurpassable politeness, delicacy, sweet temper, hospitality, consideration, and with unsurpassable respect for the privacy daily enforced upon me by the nature of my avocation here and the state of my health. This testimony, so long as I live, and so long as my descendants have any legal right in my books, I shall cause to be republished as an appendix to every copy of those two books of mine in which I have referred to America. And this I will do and cause to be done, not in mere love and thankfulness, but because I regard it as an act of plain justice and honour.'

The time for Mr. Dickens' departure was now close at hand. His last reading was given at the Steinway Hall, on the ensuing Monday evening. The task finished, he was about to retire, but a tremendous burst of applause stopped him. He knew what his audience wanted—a few words, a parting greeting before saying good-bye. Their illustrious visitor did not disappoint them. 'The shadow of one word has impended over me this evening,' said Mr. Dickens, 'and the time has come at length when the shadow must fall. It is but a very short one, but the weight of such things is not measured by their length, and two much shorter words express the round of our human existance. When I was reading *David Copperfield* a few evenings since, I felt there was more than usual significance in the words of

Peggotty, "My future life lies over the sea." ... The relations which have been set up between us must now be broken for ever. Be assured, however, that you will not pass from my mind. I shall often realize you as I see you now, equally by my winter fire and in the green English summer weather. I shall never recall you as a mere public audience, but rather as a host of personal friends, and ever with the greatest gratitude, tenderness, and consideration. Ladies and gentlemen, I beg to bid you farewell. God bless you, and God bless the land in which I leave you!'

The sums gained by his last American readings were very large. His agent Dolby was paid in commission about £2888, the commission received by Messrs. Ticknor & Fields was £1000, besides five per cent. on Boston receipts. The preliminary expenses for this series of readings were £614, the expenses in America amounted to £13,000. His own profits were within a hundred or so of £19,000; united to his English receipts, he had thus gained £33,000 in two years. Towards the close of 1868 Dickens began his series of 'Farewell Readings,' which were previously settled to take place in the chief towns of England, Ireland, and Scotland. When in Liverpool in 1869, he was entertained at a splendid banquet in the St. George's Hall, the Mayor presiding. The number of ladies and gentlemen who sat down to dinner was upwards of seven hundred. In allusion to a remark in Lord Houghton's speech, that had he sought Parliamentary honours, he might have done good service to his country, he said: 'When I first took literature as my profession in England, I calmly resolved within myself that, whether I succeeded or whether I failed, literature should be my sole profession. It appeared to me at that time that it was not so well understood in England as it was in other countries that literature was a dignified profession, by which any man might stand or fall. I made a compact with myself that in my person literature should stand, and by itself, of itself, and for itself; and there is no consideration on earth which would induce me to break that bargain.' When on this course of readings, his health broke down, and he was obliged to retire from the platform for a time in order to recruit. He was able on the 27th August to attend the dinner given by the London Rowing Club to the crews of Oxford and Harvard Universities; and on the 27th of September following, he delivered the annual address at the commencement of the winter session of the Birmingham and Midland Institute. In order to avoid the frequent journeys to and from Gad's Hill, for six months he rented a house in Hyde Park Place, where a considerable portion of his unfinished novel *Edwin Drood* was written. His last reading took place at St. James' Hall, London, on the 15th of March 1870, and consisted of the 'Christmas Carol,' and 'The Trial' from *Pickwick*. The hall was crowded in every part, having been filled as soon as the doors were open, and thousands were unable to find admittance. His reading was even more spirited and energetic than ever, and his voice was clear to the last. When the applause at the conclusion had subsided, Dickens spoke as follows:—' Ladies and gentlemen,—It would be worse than idle, for it would be hypocritical and unfeeling, if I were to disguise that I close this episode in my life with feelings of very considerable pain. For some fifteen years, in this hall and in many kindred places, I have had the honour of presenting my own cherished ideas before you for your recognition, and, in closely observing your reception of them, have enjoyed an amount of artistic delight and instruction which, perhaps, is given to few men to know. In this task, and in every other I have ever undertaken, as a faithful servant of the public, always imbued with a sense of duty to them, and always striving to do his best, I have been uniformly cheered by the readiest response, the most generous sympathy, and the most stimulating support. Nevertheless I have thought it well, at the full flood-tide of your favour, to retire upon those older associations between us, which date from much farther back than these, and henceforth to devote myself exclusively to the art that first brought us together. Ladies and gentlemen, in but two short weeks from this time I hope that you may enter, in your own homes, on a new series of readings, at which my assistance will be indispensable;[1] but from these garish lights I vanish now for evermore, with a heartfelt, grateful, respectful, and affectionate farewell.'

Her Majesty the Queen, interested in his life and work, requested him to attend her at Buckingham Palace on 9th April 1870. He was introduced to Her Majesty by his friend Mr. Arthur Helps, Clerk of the Privy Council. The interview lasted for some time, in the course of which Her Majesty expressed her admiration of and interest in his works, and presented him on parting with a copy of *Our Life in the Highlands*, with this autograph inscription, ' Victoria R., to Charles Dickens.' He sent her in return an edition of his collected works, which the Queen graciously placed in her own private library. After his death it became known that the Queen had been anxious to bestow upon him some distinction in keeping with his views and tastes, and

[1] Alluding to the forthcoming serial story of *Edwin Drood*.

it is just possible that he might have been asked to accept a place in her Privy Council. It was noticed that about this time he appeared rather more in society than usual, although continuing to complain that he was unwell. His last public appearances were in April, and his last public speech was a graceful tribute, at the Academy dinner, to his friend Daniel Maclise.

On the 7th of May he read the fifth number of *Edwin Drood* to John Forster. About this time he dined with Mr. Motley, the American minister, met Mr. Disraeli at Lord Stanhope's, and also appeared at breakfast with Mr. Gladstone. He had an invitation for the 17th of the month to attend the Queen's ball, but this he was unable to do owing to disablement. On the 16th he wrote to Mr. Forster : ' I am sorry to report that, in the old preposterous endeavour to dine at preposterous hours and preposterous places, I have been pulled up with a sharp attack in my foot. And serve me right. I hope to get the better of it soon, but I fear I must not think of dining with you on Friday. I have cancelled everything in the dining way for this week, and that is a very small precaution after the horrible pain I have had and the remedies I have taken.' He declined to attend the General Theatrical Fund dinner, when the Prince of Wales was to preside, but he dined with Lord Houghton a week later. On the 30th May he quitted London for Gad's Hill, where he confined his attention closely to his novel *Edwin Drood*, which was in progress. He was observed now to have a very wearied appearance. On Monday, 6th June, he was out with his dogs for the last time, when he walked into Rochester. On the following day he drove out. The 8th of June was spent in writing in the châlet in the garden, uninterrupted save for luncheon. He was late in leaving the châlet; dinner was ordered for six o'clock, but before that time he wrote several letters, one to Mr. Charles Kent, arranging to see him in London next day. At dinner his sister-in-law, Miss Hogarth, noticed with pain a troubled expression in his face. 'For an hour,' he said, 'he had been very ill, but he did not wish the dinner to be interrupted.' These were the last coherent words said to have been uttered by him. His talk continued to be rambling, and in attempting to rise, his sister-in-law's help alone prevented him from falling where he stood. Endeavouring to get him to the sofa, he sank heavily to the ground, falling on his left side. 'On the ground' were the last words spoken by him. His family were telegraphed for, and medical aid was called in; but the case was hopeless, and he died on the evening of Thursday, 9th June, having lived four months beyond his fifty-eighth year. The immediate cause of death was from effusion of blood on the brain, brought on by overwork. The public journals, from the *Times* onwards, gave expression to their feeling at the loss which had been sustained. The Queen telegraphed her regret from Balmoral, where she had been staying. A grave in Westminster Abbey was offered for the deceased; and although he would himself have preferred to have found his last resting-place at Rochester, the proposal was accepted, and the funeral, strictly private, took place on Tuesday, 14th June. The stone placed upon his grave is inscribed—

CHARLES DICKENS.

BORN FEBRUARY THE SEVENTH 1812.

DIED JUNE THE NINTH 1870.

His will had just been completed seven days before he was struck down. By a codicil to the will, his interest in *All the Year Round* was left to his acting editor and eldest son, with other private instructions for their guidance in the conduct of the journal.

The funeral sermon was preached by Dean Stanley, on Sunday, the 19th June. Amongst the unnoted thousands present were Thomas Carlyle and Alfred Tennyson. The text of the day was the verses in the 15th and 16th chapters of Luke—the parable of the rich man and Lazarus. In the course of his sermon the Dean remarked :—

'It is said to have been the distinguishing glory of a famous Spanish saint that she was the advocate of the absent. That is precisely the advocacy of this divine parable, and of those modern parables which most represent its spirit—the advocacy, namely, of the poor, the absent, the neglected, of the weaker side, whom, not seeing, we are tempted to forget. It was the part of him whom we have lost to make the rich man, faring sumptuously every day, not fail to see the presence of the poor man at his gate. The suffering inmates of our workhouses; the neglected children in the dens and caves of this great city; the starved ill-used boys in remote schools, far from the observation of men,—these all felt a new ray of sunshine poured into their dark prisons, and a new interest awakened in their forlorn and desolate lot, because an unknown friend had pleaded their cause with a voice that rang through the palaces of the great as well as through the cottages of the poor. In his pages, with gaunt figures and hollow voices, they were made to stand and speak before those who had before hardly dreamed of their existence. But was it mere compassion which this created? The same master hand which drew the sorrows of the English poor drew also the picture of the unselfishness, the kindness, the courageous patience, and the tender thoughtfulness that lie concealed under many a coarse exterior, and are to be found in many a degraded home. When the little workhouse boy wins his way, pure and undefiled, through the mass of wickedness around him; when the little orphan girl, who brings thoughts of heaven into the hearts of all around her, is as the very gift of God to the old man who sheltered her life,—these are scenes which no human being can read without being the better of it. He laboured to teach us that there is even in the worst of mankind a soul of goodness—a soul worth revealing, worth reclaiming, worth regenerating. He laboured to teach the rich and educated how this better side was to be found, even in the most neglected Lazarus, and to tell the poor no less to respect this better part of themselves—to remember that they also have a calling to be good and great, if they will but hear it.

'There is one more thought that arises on this occasion. As, in the parable, we are forcibly impressed with the awful solemnity of the other world, so on this day a feeling rises in us before which the most brilliant powers of genius and the most lively sallies of wit wax faint. When, on Tuesday last, we stood beside that open grave, in the still, deep silence of the summer morning, in the midst of this vast solitary space, broken only by that small band of fourteen mourners, it was impossible not to feel that there is something more sacred than any worldly glory, however bright, or than any mausoleum, however mighty; and that is the return of the human soul into the hands of its Maker. Many, many are the feet that have trodden, and will tread, the consecrated ground around his grave. Many, many are the hearts which, both in the Old

World and the New, are drawn towards it as towards the resting-place of a dear personal friend. Many are the flowers that have been strewn—many the tears that have been shed—by the grateful affection of the poor that have cried, of the fatherless, and of those that have none to help them. May I speak to them a few sacred words, that will come perhaps with a new meaning and a deeper force, because they come from the lips of their lost friend, because they are the most solemn utterances of lips now closed for ever in the grave? They are extracted from the will of Charles Dickens, dated 12th May 1869, and will now be heard by many for the first time. After the most emphatic injunctions respecting the inexpensive, unostentatious, and strictly private manner of his funeral,—injunctions which have been carried out to the very letter,—he thus continues:

'"*I direct that my name be inscribed in plain English letters on my tomb. I conjure my friends on no account to make me the subject of any monument, memorial, or testimonial whatever. I rest my claim to the remembrance of my country on my published works, and to the remembrance of my friends in their experience of me in addition thereto. I commit my soul to the mercy of God, through our Lord and Saviour Jesus Christ; and I exhort my dear children humbly to try to guide themselves by the teaching of the New Testament, in its broad spirit, and to put no faith in any man's narrow construction of its letter here or there.*"

'In that simple but sufficient faith he lived and died. In that simple and sufficient faith he bids you live and die. If any of you have learnt from his works the value—the eternal value—of generosity, of purity, of kindness, of unselfishness, and have learnt to show these in your own hearts and lives, then remember that these are the best monuments, memorials, and testimonials of the friend whom you have loved, and who loved with a marvellous and exceeding love his children, his country, and his fellow-men. These are monuments which he would not refuse, and which the humblest and poorest and youngest here have it in their power to raise to his memory.'

The beautiful anthem, 'When the ear heard him,' was then sung, and the remainder of the service was gone through.

THE END

MORRISON AND GIBB, PRINTERS, EDINBURGH.

HEROES

OF

INVENTION AND DISCOVERY.

MORRISON AND GIBB, PRINTERS, EDINBURGH.

GEORGE STEPHENSON

HEROES

OF

INVENTION AND DISCOVERY:

*LIVES OF EMINENT INVENTORS AND
PIONEERS IN SCIENCE.*

SELECTED BY THE EDITOR OF

'RISEN BY PERSEVERANCE; OR, LIVES OF SELF-MADE MEN," "THE
ENGLISH ESSAYISTS," "TREASURY OF MODERN BIOGRAPHY," ETC.

EDINBURGH:
W. P. NIMMO, HAY, & MITCHELL

PREFATORY NOTE

THIS book, like its companions in the same series, has been produced by the Publishers with a view of providing biographical reading of a wholesome and instructive character, free from sectarian bias of any kind. The importance of invention and discovery in all moral, commercial, and intellectual progress will be readily conceded by every one, and little apology is needed now in presenting examples of some of those eminent in these departments. The articles on James Watt, Robert Boyle, and Sir Humphrey Davy are drawn from a well-known book, "The Pursuit of Knowledge under Difficulties," by the late Professor Craik; London, 1830; with some slight exceptions, the other articles composing the bulk of the book are selected from copyright material placed at the disposal of the Editor for use in the present volume.

CONTENTS.

	PAGE
ROBERT BOYLE,	1
JAMES WATT,	20
SIR HUMPHREY DAVY,	50
GEORGE STEPHENSON,	69
SIR JAMES Y. SIMPSON,	121
GALLERY OF GREAT INVENTORS AND DISCOVERERS—	
ROGER BACON,	146
WILLIAM LEE,	151
MARQUIS OF WORCESTER,	154
PRINCE RUPERT,	156
SIR SAMUEL MORLAND,	158
JOHN FLAMSTEAD,	159
JOHN HARRISON,	161
GEORGE GRAHAM,	162
JAMES FERGUSON,	164
MATTHEW BOULTON,	167
JOSEPH BLACK,	170
JOSEPH PRIESTLEY,	173
JAMES HARGREAVES,	177
JOSIAH WEDGWOOD,	178
HENRY CORT,	181
SAMUEL CROMPTON,	184
HENRY BELL,	189
SIR DAVID BREWSTER,	190
CHARLES BABBAGE,	194
HENRY BESSEMER,	196
JOHN ERICSSON,	205
THOMAS ALVA EDISON,	212

HEROES OF INVENTION AND DISCOVERY.

ROBERT BOYLE.

PERHAPS the best example we can adduce of the manner in which wealth may be made subservient by its possessor, not only to the acquisition of knowledge, but also to its diffusion and improvement, is that of our celebrated countryman The Honourable ROBERT BOYLE. Boyle was borne at Lismore, in Ireland, in 1627, and was the seventh and youngest son of Richard, the first Earl of Cork, commonly called the Great Earl. The first advantage which he derived from the wealth and station of his father was an excellent education. After having enjoyed the instructions of a domestic tutor, he was sent, at an early age, to Eton. But his inclination, from the first, seems to have led him to the study of things, rather than of words. He remained at Eton only four years, "in the last of which," according to his own statement, in an account which he has given us of his early life, " he forgot much of that Latin he had got, for he was so addicted to more solid parts of knowledge, that he hated the study of bare words naturally, as something that relished too much of pedantry, to consort with his disposition and designs." In reference to what is here insinuated, in disparagement of the study of languages merely as such, we may just remark that the observation is, perhaps, not quite so profound as it is plausible. So long as one mind differs from another, there will always be much difference of sentiment as to the comparative claims upon our regard of that, on the one hand, which addresses itself principally to the taste or the imagination, and that, on the other, which makes its appeal to the understanding only. But it is, at any rate, to be remembered that, in confining the epithet useful, as is commonly done, to the latter, it is intended to describe it as the useful only pre-eminently, and not exclusively. The agreeable or the graceful is plainly also useful. The study of language and style, therefore, cannot with any propriety be denounced as a mere waste of time; but, on the contrary, is well fitted to become to the mind a source both of enjoyment and of power. So great, indeed, is the influence of diction upon the common feelings of mankind, that no literary work, it may be safely asserted, has ever acquired a permanent reputation and popularity, or, in other words, produced any wide and enduring effect, which was not distinguished by the graces of its style. Their deficiency, in this respect, has been at least one of the causes of the comparative oblivion into which Mr. Boyle's own writings have fallen, and, doubtless, weakened the efficacy of such of them as aimed at anything beyond a bare statement of facts, even in his own day. It was this especially which exposed some of his moral lucubrations to Swift's annihilating ridicule.

On being brought home from Eton, Boyle, who was his father's favourite son, was placed under the care of a neighbouring clergyman, who, instructing him, he says, "both with care and civility, soon brought him to renew his first acquaintance with the Roman tongue, and to improve it so far that in that language he could readily enough express himself in prose, and began to be no dull proficient in the poetic strain." "Although, however," he adds, "naturally addicted to poetry, he forbore, in after-life, to cultivate his talent for that species of composition, because, in his travels, having by discontinuance forgot much of the Latin tongue, he afterwards never could find time to redeem his losses by a serious study of the ancient poets." From all this it is evident that the natural bent of his mind did not incline him very strongly to classical studies; and as, for the most obviously wise purposes, there has been established among men a diversity of intellectual endowments and tendencies, and every mind is most efficient when it is employed most in accordance with its natural dispositions and predilections, it was just as well that the course of his education was now changed. In his eleventh year he and one of his brothers were put under the charge of a Mr. Marcombes, a French gentleman, and sent to travel on the Continent. In the narrative of his early life, in which he designates himself by the name of Philoretus, Mr. Boyle has left us an account of his travelling tutor. "He was a man," says he, "whose gait, his mien, and outside, had very much of his nation, having been divers years a traveller and a soldier; he was well fashioned, and very well knew what belonged to a gentleman. His natural were much better than his acquired parts, though divers of the latter he possessed, though not in an eminent, yet in a competent degree. Scholarship he wanted not, having in his greener years been a professed student in divinity; but he was much less read in books than men, and hated pedantry as much as any of the seven deadly sins. . . . Before company he was always very civil to his pupils, apt to eclipse their failings, and set off their good qualities to the best advantage. But in his private conversation he was cynically disposed, and a very nice critic both of words and men; which humour he used to exercise so freely with Philoretus, that at last he forced him to a very cautious and considerate way of expressing himself, which after turned to his no small advantage. The worst quality he had was his choler, to excesses of which he was excessively prone; and that being the only passion to which Philoretus was much observed to be inclined, his desire to shun clashing with his governor, and his accustomedness to bear the sudden sallies of his impetuous humour, taught our youth so to subdue that passion in himself, that he was soon able to govern it habitually and with ease."

Under the guidance of this gentleman, who, although not much fitted, apparently, to make his pupils profound scholars, or even to imbue them with a taste for elegant literature, was, probably, very well qualified both to direct their powers of observation, and to superintend and assist the general growth of their minds at this early age, the two brothers passed through France to Geneva, where they continued some time studying rhetoric, logic, mathematics, and political geography, to which were added the accomplishments of fencing and dancing. "His recreations during his stay at Geneva," says Mr. Boyle of himself, "were sometimes mall, tennis (a sport he ever passionately loved), and, above all, the reading of romances, whose perusal did not only extremely divert him, but (assisted by a total discontinuance of the English tongue) in a short time taught him a skill in French somewhat unusual to strangers." The party afterwards set off for Italy; and, after visiting Venice and other places, proceeded to Florence, where they spent the winter.

While residing here Mr. Boyle made himself master of the Italian language. But another acquisition, for which he was indebted to his visit to Florence, probably influenced to a greater extent the future course of his pursuits; we mean the knowledge he obtained of the then recent astronomical discoveries of Galileo. This great philosopher died in the neighbourhood of Florence, in the beginning of the year 1642, while Boyle and his brother were pursuing their studies in that city. The young Englishman, who was himself destined to acquire so high a reputation by his experiments in various departments of physical science, some of them the same which Galileo had cultivated, probably never even beheld his illustrious precursor; but we cannot tell how much of Boyle's love of experimental inquiry, and his ambition to distinguish himself in that field, may have been caught from this, his accidental residence in early life in a place where the renown of Galileo and his discoveries must have been on the lips of all.

Boyle returned to England in 1644. Although he was yet only in his eighteenth year, he seems to have thought that his education had been long enough under the direction of others, and he resolved, therefore, for the future to be his own instructor. Accordingly, his father being dead, he retired to an estate which had been left him in Dorsetshire, and gave himself up, we are told, for five years to the study principally of natural philosophy and chemistry. His literary and moral studies, however, it would appear, were not altogether suspended during this time. In a letter written by him from his retirement to his old tutor, Mr. Marcombes, we find him mentioning, as also among his occupations, the composing of essays in prose and verse, and the study of ethics, "wherein," says he, "of late I have been very conversant, and desirous to call them from the brain down into the breast, and from the school to the house."

These details do not, like many of those we have given in other parts of our work, exhibit to us the ardent lover of knowledge, beset with impediments at every step, in his pursuit of the object on which he has placed his affections, and having little or nothing to sustain him under the struggle, except the unconquerable strength of the passion with which his heart is filled. On the contrary, we have here a young man who has enjoyed from his birth upwards every facility for the improvement of his mind, and is now surrounded with all the conveniences he could desire, for a life of the most various and excursive study. A happy and enviable lot! Yet by how few of those to whom it has been granted, as well as to him of whom we are now speaking, have its advantages been used as they were by him! The truth is, that if the mind be not in love with knowledge, no mere outward advantages will enable any one to make much progress in the pursuit of it; while with this love for it, all the difficulties which the unkindness of fortune can throw in the way of its acquisition may be overcome. The examples frequently recorded of many a successful struggle with such difficulties in their most collected and formidable strength, sufficiently warrant us to hold out this encouragement to all.

In the same letter to Mr. Marcombes, which we have just quoted, we find Boyle making mention, for the first time, of what he calls "our new Philosophical or Invisible College," some of the leading members of which, he informs his correspondent, occasionally honoured him with their company at his house. By this *Invisible College,* he undoubtedly means that association of learned individuals who began about this period to assemble together in London for the purposes of scientific discussion, and whose meetings formed the germ of the Royal Society. According to the account given in a letter written many years after by Dr. Wallis, another member of the club, to his friend Dr. Thomas Smith, it appears that these meetings first began to be held in London, on a certain day in every week, about the year 1645. Mr. Boyle's name does not occur in the list of original members given by Dr. Wallis; but he professes to mention only several of the number. There can be no doubt that Boyle joined them soon after the formation of the association. According to Dr. Wallis, the meetings were first suggested by a Mr. Theodore Haak, whom he describes as a German of the Palatinate, then resident in London. They used to be held sometimes in Wood Street, at the house of Dr. Goddard, the eminent physician, who kept an operator for grinding glasses for telescopes and microscopes; sometimes at another house in Cheapside; and sometimes in Gresham College, to which several of the members were attached. The subjects of inquiry and discussion are stated to have embraced everything relating to "physic, anatomy, geometry, astronomy, navigation, magnetics, chemics, mechanics, and natural experiments," whatever, in short, belonged to what was then called "the new or experimental philosophy." In course of time several of the members of the association were removed to Oxford; and they began at last to meet by themselves in that city, while the others continued their meetings in London. The Oxford meetings began to be regularly held about the year 1649. In 1654 Mr. Boyle took up his residence at Oxford, probably induced, in great part, by the circumstance of so many of his philosophical friends being now there, and engaged together in the same inquiries with himself. The Oxford associates, according to Dr. Wallis, met first in the apartments of Dr. Petty (afterwards the celebrated Sir William Petty, the ancestor of the Marquis of Lansdowne), who lodged, it seems, in the house of an apothecary, whose store of drugs was found convenient for their experiments. On Dr. Petty going to Ireland, they next met, the narrative proceeds, "(though not so constantly) at the lodgings of Dr. Wilkins, then warden of Wadham College; and, after his removal to Trinity College, in Cambridge, at the lodgings of the honourable Mr. Robert Boyle, then resident for divers years in Oxford." Boyle, indeed, continued to reside in this city till the year 1668. Meanwhile, in 1663, three years after the Restoration, the members of the London club were incorporated under the title of the Royal Society.

It was during his residence at Oxford that Boyle made some of the principal discoveries with which his name is connected. In particular, it was here that he prosecuted those experiments upon the mechanical properties of the air, by which he first made himself generally known to the public, and the results of which rank among the most important of his contributions to natural science. The first account which he published of these experiments appeared at Oxford in 1660, under the title of "New Experiments Physico-Mechanical, touching the spring of the air and its effects." The work is in the form of letters to his nephew, Viscount Dungarvon, the son of the Earl of Cork, which are dated in December, 1659. It may be not unnaturally supposed that Boyle's attention was first directed to the subject of Pneumatics, when he was engaged at Florence in making himself acquainted with the discoveries of Galileo, whose experiments first introduced anything like science into that department of inquiry. He states, himself, in his first letter to his nephew, that he had some years before heard of a book, by the Jesuit Schottus, giving an account of a contrivance, by which

Otto Guericke, Consul of Magdeburg, had succeeded in emptying glass vessels of their contained air, by sucking it out at the mouth of the vessel, plunged under water. He alludes here to Guericke's famous invention of the instrument now commonly called the air-pump. This ingenious and ardent cultivator of science, who was borne in Magdeburg, in Saxony, in the beginning of the seventeenth century, in his original attempts to produce a vacuum, used first to fill his vessel with water, which he then sucked out by a common pump, taking care, of course, that no air entered to replace the liquid. This method was probably suggested to Guericke by Torricelli's beautiful experiment with the barometrical tube, the vacuum produced in the upper part of which, by the descent of the mercury, has been called from him the Torricellian vacuum. It was by first filling it with water that Guericke expelled the air from the copper globe, the two closely-fitting hemispheres comprising which six horses were then unable to pull asunder, although held together by nothing more than the pressure of the external atmosphere.

This curious proof of the force, or weight of the air, which was exhibited before the Emperor Ferdinand III., in 1654, is commonly referred to by the name of the experiment of the Magdeburg hemispheres. Guericke, however, afterwards adopted another method of exhausting a vessel of its contained air, which could be applied more generally than the one he had first employed. This consisted in at once pumping out the air itself. The principle of the contrivance which he used for that purpose will be understood from the following explanation. If we suppose a barrel of perfectly equal bore throughout, and having in it a closely-fitting plug or piston, to have been inserted in the mouth of the vessel, it is evident that, when this piston was drawn up from the bottom to the top of the barrel, it would carry along with it all the air that had previously filled the space through which it had passed. Now were air, like water, possessed of little or no expansive force, this space, after being thus deprived of its contents, would have remained empty, and there would have been an end of the experiment. But in consequence of the extraordinary elasticity of the element in question, no sooner would its original air be lifted by the piston out of the barrel, than a portion of that in the vessel beyond the piston would flow out to occupy its place. The vessel and the barrel together would now, therefore, be filled by the same quantity of air which had originally been contained in the first alone, and which would consequently be diminished in density just in proportion to the enlargement of the space which it occupied. But although so much of the air to be extracted had thus got again into the barrel, there would still at this point have been an end of the experiment, if no way could have been found of pushing back the piston for another draught, without forcing also the air beyond it into the vessel again, and thus merely restoring matters to the state in which they were at the commencement of the operation. But here Guericke was provided with an ingenious contrivance—that of the valve; the idea of applying which he borrowed, no doubt, from the common water-pump, in which it had been long used. A valve, which, simple as it is, is one of the most useful and indeed indispensable of mechanical contrivances, is, as most persons know, merely a flap, or lid, moving on a hinge, which, covering an orifice, closes it, of course, against whatever attempts to pass through from behind itself (a force bearing upon it from thence evidently only shutting it closer), while it gives way to and permits the passage of whatever comes in the opposite direction. Now Guericke, in his machine, had two of these valves, one covering a hole in the piston, another covering the mouth of the vessel where the barrel was inserted; and both opening outwards. In consequence of this arrangement, when the piston, after having been drawn out, as we have already described, was again pushed back, the air in the barrel was prevented from getting back into the vessel by the farther valve, now shut against it, while it was at the same time provided with an easy means of escape by the other, through which, accordingly, it passed away. Here, then, was one barrelful of the air in the vessel dislodged; and the same process had only to be repeated a sufficient number of times in order to extract as much more as was desired. The quantity, however, removed every time was, of course, always becoming less; for, although it filled the same space, it was more attenuated.

The principle, therefore, upon which the first air-pump was constructed was the expansibility of the air, which the inventor was enabled to take advantage of through means of the valve. These two things, in fact, constitute the air-pump; and whatever improvements have been since introduced in the construction of the machine have gone only to make the working of it more convenient and effective. In this latter respect the defects of Guericke's apparatus, as might be expected, were considerable. Among others, with which it was chargeable, it required the continual labour of two men for several hours at the pump to exhaust the air from a vessel of only moderate size; the precautions which Guericke used to prevent the intrusion of air from without, between the piston and the sides of the barrel, during the working of the machine, were both imperfect for that purpose, and greatly added to the difficulties and incommodiousness of the operation; and, above all, from the vessel employed being a round globe, without any other mouth or opening than the narrow one in which the pump was inserted, things could not be conveyed into it, nor, consequently, any experiments made in that vacuum which had been obtained. Boyle, who says that he had himself thought of something like an air-pump before he heard of Guericke's invention, applied himself, in the first place, to the remedying of these defects in the original instrument, and succeeded in rendering it considerably more convenient and useful. At the time when he began to give his attention to this subject, he had Robert Hooke, who afterwards attained a distinguished name in science, residing with him as an assistant in his experiments; and it was Hooke, he says, who suggested to him the first improvements in Guericke's machine. These, which could not easily be made intelligible by any mere description, and which, besides, have long since given way to still more commodious modifications of the apparatus, so that they possess now but little interest, enabled Boyle and his friends to carry their experiments with the new instrument much farther than had been done by the Consul of Magdeburg. But, indeed, Boyle himself did not long continue to use the air-pump which he describes in this first publication. In the second part of his Physico-Mechanical Experiments he describes one of a new construction; and, in the third part of the same work, one still farther improved. This last, which is supposed to have been also of Hooke's contrivance, had two barrels moved by the same pinion-wheel, which depressed the one while it elevated the other, and thus did twice as much work as before in the same time. The air-pump has been greatly improved since the time of Boyle by the Abbé Nollet, Gravesande, Smeaton, Prince, Cuthbertson, and others.

By his experiments with this machine Boyle made several important discoveries with regard to the air, the principal of which he details in the three successive parts of the work we have mentioned. Having given so commodious a form and position to the vessel out of which the air was to be extracted (which, after him, has been generally called the receiver, a name, he says, first bestowed upon it by the glassmen), that he could easily introduce into it anything which he wished to make the subject of an experiment, he found that neither flame would burn nor animals live in a vacuum, and hence he inferred the necessity of the presence of air both to combustion and animal life. Even a fish, immersed in water, he proved, would not live in an exhausted receiver. Flame and animal life, he showed, were also both soon extinguished in any confined portion of air, however dense, although not so soon in a given bulk of dense as of rarefied air; nor was this, as had been supposed, owing to any exhalation of heat from the animal body or the flame, for the same thing took place when they were kept in the most intense cold, by being surrounded with a frigorific mixture. What he chiefly sought to demonstrate, however, by the air-pump was, the extraordinary elasticity, or spring, as he called it, of the air. It is evident, from the account that has been given of the principle of this machine, that, if the pump be worked ever so long, it never can produce in the receiver a strictly perfect vacuum; for the air expelled from the barrel by the last descent of the piston must always be merely a portion of a certain quantity, the rest of which will be in the receiver. The receiver, in truth, after the last stroke of the piston, is as full of air as it was at first; only that by which it is now filled is so much rarefied and reduced in quantity, although it occupies the same space as before, that it may be considered as, for most practical purposes, annihilated. Still a certain quantity, as we have said, remains, be it ever so small; and this quantity continues, just as at first, to be diffused over the whole space within the receiver. From this circumstance Boyle deduced some striking evidences of what seems to be the almost indefinite expansibility of the air. He at last actually dilated a portion of air to such a degree that it filled, he calculated, 13,679 times its natural space, or that which it occupied as part of the common atmosphere. But the usual density of the atmosphere is very far from being the greatest to which the air may be raised. It is evident that, if the two valves of the air-pump we have already described be made to open inwards instead of outwards, the effect of every stroke of the piston will be, not to extract air from the receiver, but to force an additional quantity into it. In that form, accordingly, the machine is called a forcing-pump, and is used for the purpose of condensing air, or compressing a quantity of it into the smallest possible space. Boyle succeeded, by this method, in forcing into his receiver forty times its natural quantity. But the condensation of the air has been carried much further since his time. Dr. Hales compressed into a certain space 1522 times the natural quantity, which in this state had nearly twice the density, or, in other words, was nearly twice as heavy as the same bulk of water. Of the air thus condensed by Dr. Hales, therefore, the same space actually contained above twenty millions of times the quantity which it would have done of that dilated to the highest degree by Mr. Boyle. How far do these experiments carry us beyond the knowledge of Aristotle, who held that the air, if rarefied so as to fill ten times its usual space, would become fire!

On leaving Oxford, in 1668, Boyle came to London, and

here he continued to reside during the remainder of his life. Up to this time his attendance at the meetings of the Royal Society had been only occasional, but he was now seldom absent. Science, indeed, was as much the occupation of his life as if it had been literally his business or profession. No temptations could seduce him away from his philosophical pursuits. Belonging, as he did, to one of the most powerful families in the kingdom—having no fewer than four brothers in the Irish peerage, and one in the English,—the highest honours of the State were open to his ambition if he would have accepted of them. But so pure was his love of science and learning, and, with all his acquirements, so great his modesty, that he steadily declined even those worldly distinctions which might be said to lie strictly within the sphere of his pursuits. He was zealously attached to the cause of religion, in support of which he wrote and published several treatises; but he would not enter the Church, although pressed to do so by the king, or even accept of any office in the universities, under the conviction that he should more effectually serve the interests both of religion and learning by avoiding everything which might give him the appearance of being their hired or interested advocate. He preferred other modes of showing his attachment, in which his wealth and station enabled him to do what was not in the power of others. He allowed himself to be placed at the head of associations for the prosecution of those objects which he had so much at heart; he contributed to them his time, his exertions, and his money; he printed, at his own expense, several editions of the Scriptures in foreign languages for gratuitous distribution; if learned men were in pecuniary difficulties, his purse was open to their relief. And, as for his own labours, no pay could have made them more zealous or more incessant. From his boyhood till his death he may be said to have been almost constantly occupied in making philosophical experiments; collecting and ascertaining facts in natural science; inventing or improving instruments for the examination of nature; maintaining a regular correspondence with scientific men in all parts of Europe; receiving the daily visits of great numbers of the learned both of his own and other countries; perusing and studying not only all the new works that appeared in the large and rapidly widening department of natural history and mathematical and experimental physics, including medicine, anatomy, chemistry, geography, &c., but many others relating especially to theology and Oriental literature; and lastly, writing so profusely upon all these subjects, that those of his works alone which have been preserved and collected, independently of many others that are lost, fill, in one edition, six large quarto volumes. So vast an amount of literary performance, from a man who was at the same time so much of a public character, and gave so considerable a portion of his time to the service of others, shows strikingly what may be done by industry, perseverance, and such a method of life as never suffers an hour of the day to run to waste.

In this last particular, indeed, the example of Mr. Boyle well deserves to be added to those of the other distinguished men in this department. Of his time he was, from his earliest years, the most rigid economist, and he preserved that good habit to the last. Dr. Dent, in a letter to Dr. Wotton, tells us that "his brother, afterwards Lord Shannon (who accompanied him on his continental tour with Mr. Marcombes), used to say that even then he would never lose any vacant time; for, if they were upon the road, and walking down a hill, or in a rough way, he would read all the way; and when they came at night to their inn he would still be studying till supper, and frequently propose such difficulties as he met with in his reading to his governor." The following naïve statement, too, which we find in an unfinished essay on a theological subject, which he left behind him in manuscript, and of which Dr. Birch, the editor of his collected works, has printed a part, may serve to show the diligence with which he prosecuted his severer studies, even amidst all sorts of interruptions. "It is true," he writes, "that a solid knowledge of that mysterious language" (it is his acquisition of the Hebrew tongue to which he refers) "is somewhat difficult, but not so difficult but that so slow a proficient as I could, in less than a year, of which not the least part was usurped by frequent sicknesses and journeys, by furnaces, and by (which is none of the modestest thieves of time) the conversation of young ladies, make a not inconsiderable progress towards the understanding of both Testaments in both their originals." But the life of active and incesssant occupation which he led, even in his declining years, is best depicted in another curious document which Dr. Birch has preserved. A few years before his death he was urged to accept the office of President of the Royal Society, of which he had so long been one of the most active and valuable members, and the Transactions of which he had enriched by many papers of great interest; but he declined the honour on the score of his growing infirmities. About this time he also published an advertisement, addressed to his friends and acquaintances, in which he begins by remarking "that he has, by some unlucky accidents, had many of his writings corroded here and there, or otherwise so maimed" (this is a specimen of the pedantic mode of expression of which Boyle was too fond), "that without he himself fill up the *lacunæ* out of his memory or invention, they will not be intelligible." He then goes on to allege his age and his ill health as reasons for immediately setting about the arrangement of his papers, and to state that his physician and his best friends have "pressingly advised him against speaking daily with so many persons as are wont to visit him;" representing it as that which must "disable him for holding out long." He therefore intimates that he means in future to reserve two days of the week to himself, during which, "unless upon occasions very extraordinary," he must decline seeing either his friends or strangers, "that he may have some time both to recruit his spirits, to range his papers, and fill up the *lacunæ* of them, and to take some care of his affairs in Ireland, which are very much disordered, and have their face often changed by the public disorders there." He at the same time ordered a board to be placed over his door, giving notice when he did and when he did not receive visits.

Nothing can set in a stronger light than this the celebrity and public importance to which he had attained. His reputation, indeed, had spread over Europe; and he was the principal object of attraction to all scientific strangers who visited the English metropolis. Living, as it was his fortune to do, at what may be called only the dawn of modern science, Boyle perhaps made no discovery which the researches of succeeding investigators in the same department have not long ere now gone far beyond. But his experiments, and the immense number of facts which he collected and recorded, undoubtedly led the way to many of the most brilliant results by which, since his day, the study of nature has been crowned. Above all, he deserves to be regarded as one of the principal founders of our modern chemistry. That science, before his time, was little better than a collection of dogmas, addressing themselves rather to the implicit faith of men than either to their experience or their reason. These venerable articles of belief he showed the necessity of examining, in reference to their agreement with the ascertained facts of nature; and, by bringing them to this test, exposed the falsehood of many of them. His successors have only had to contribute each his share in building up the new system; he had also to overthrow the old one.

Mr. Boyle died, at the age of sixty-four, in 1691. The experimental science of modern times never had a more devoted follower; and he claims to be recorded as having not only given us an illustrious example of the ardent pursuit of philosophy in a man of rank, but as having dedicated to its promotion the whole advantages of which his station and fortune put him in possession, with a zealous liberality that has scarcely been surpassed or equalled. Other wealthy patrons of literature and science have satisfied themselves with giving merely their money, and the *éclat* of their favourable regard to the cause which they professed to take under their protection; but he spent his life in the active service of philosophy, and was not more the encourager and supporter of all good works done in that name than a fellow-labourer with those who performed them. For the long period during which he was, in this country, the chief patron of science, he was also and equally its chief cultivator and extender. He gave to it not only his name, his influence, and his fortune, but his whole time, faculties, and exertions.

JAMES WATT

ALL the inventions and improvements of recent times, if measured by their effects upon the condition of society, sink into insignificance, when compared with the extraordinary results which have followed the employment of steam as a mechanical agent. To one individual, the illustrious JAMES WATT, the merit and honour of having first rendered it extensively available for that purpose are pre-eminently due. The force of steam, now so important an agent in mechanics, was nearly altogether overlooked until within the two last centuries. The only application of it which appears to have been made by the ancients, was in the construction of the instrument which they called the Æolipile, that is, the Ball of Æolus The Æolipile consisted of a hollow globe of metal, with a long neck, terminating in a very small orifice, which, being filled with water and placed on a fire, exhibited the steam, as it was generated by the heat, rushing with apparently great force through the narrow opening. A common tea-kettle, in fact, is a sort of Æolipile. The only use which the ancients proposed to make of this contrivance was, to apply the current of steam, as it issued from the spout, by way of a moving force—to propel for instance, the vans of a mill, or, by acting immediately upon the air, to generate a movement opposite to its own direction. But it was impossible that they should have effected any useful purpose by such methods of employing steam. Steam depends so entirely for its existence in the state of vapour upon the presence of a large quantity of heat, that it is reduced to a mist or a fluid almost immediately on coming into contact either with the atmosphere, or anything else which is colder than itself; and in this condition its expansive force is gone. The only way of employing steam with much effect, therefore, is to make it act in a close vessel. The first known writer who alludes to the prodigious energy which it exerts when thus confined, is the French engineer Solomon de Caus, who flourished in the beginning of the seventeenth century. This ingenious person, who came to England in 1612, in the train of the Elector Palatine, afterwards the son-in-law of James I., and resided here for some years, published a folio volume at Paris, in 1623, on moving forces; in which he states, that if water be sufficiently heated in a close ball of copper, the air or steam arising from it will at last burst the ball, with the noise like the going off of a petard. In another place, he actually describes a method of raising water, as he expresses it, by the aid of fire, which consists in the insertion, in the containing vessel, of a perpendicular tube, reaching nearly to its bottom, through which, he says, all the water will rise, when sufficiently heated. The agent here is the steam produced from part of the water by the heat, which, acting by its expansive force upon the rest of the water, forces it to make its escape in a jet through the tube. The supply of the water is kept up through a cock in the side of the vessel. Forty years after the publication of the work of De Caus appeared the Marquis of Worcester's famous "Century of Inventions." Of the hundred new discoveries here enumerated, the sixty-eighth is entitled "An admirable and most forcible way to drive up water by fire." As far as may be judged from the vague description which the Marquis gives us of his apparatus, it appears to have been constructed upon the same principle with that formerly proposed by De Caus; but his account of the effect produced is considerably more precise than what we find in the work of his predecessor. "I have seen the water run," says he, "like a constant fountain-stream forty feet high; one vessel of water rarefied by fire, driveth up forty of cold water." This language would imply that the Marquis had actually reduced his idea to practice; and if, as he seems to intimate, he made use of a cannon for his boiler, the experiment was probably upon a considerable scale. It is with some justice, therefore, that notwithstanding the earlier announcements in the work of the French engineer, he is generally regarded as the first person who really constructed a steam-engine.

About twenty years after this, namely in the year 1683, another of our countrymen, Sir Samuel Morland, appears to have presented a work to the French King, containing, among other projects, a method of employing steam as a mechanic power, which he expressly says he had himself invented the preceding year. The manuscript of this work is now in the British Museum; but it is remarkable that when the work, which is in French, was afterwards published by its author at Paris, in 1685, the passage about the steam-engine was omitted. Sir Samuel Morland's invention, as we find it described in his manuscript treatise, appears to have been merely a repetition of those of his predecessors, De Caus and the Marquis of Worcester; but his statement is curious as being the first in which the immense difference between the space occupied by water in its natural state and that which it occupies in the state of steam is numerically designated. The latter, he says, is about two thousand times as great as the former; which is not far from a correct account of the expansive force that steam exerts under the ordinary pressure of the atmosphere. One measure of water, it is found, in such circumstances, will produce about seventeen hundred measures of steam.

The next person whose name occurs in the history of the steam-engine, is Denis Papin, a native of France, but who spent the part of his life during which he made his principal pneumatic experiments in England. Up to this time, the reader will observe, the steam had been applied directly to the surface of the water, to raise which, in the form of a jet, by such pressure, appears to have been almost the only object contemplated by the employment of the newly-discovered power. It was Papin who first introduced a piston into the tube or cylinder which rose from the boiler. This contrivance, which forms an essential part of the common sucking-pump, is merely, as the reader probably knows, a block fitted to any tube or longitudinal cavity, so as to move freely up and down in it, yet without permitting the passage of any other substance between itself and the sides of the tube. To this block a rod is generally fixed; and it may also have a hole driven through it, to be guarded by a valve, opening upwards or downwards, according to the object in view. Long before the time of Papin it had been proposed to raise weights, or heavy bodies of any kind, by suspending them to one extremity of a handle or cross-beam attached at its other end to the rod of a piston moving in this manner in a hollow cylinder, and the descent of which, in order to produce the elevation of the weights, was to be effected by the pressure of the superincumbent atmosphere after the counterbalancing air had been by some means or other withdrawn from below it. Otto Guericke used to exhaust the lower part of the cylinder, in such an apparatus, by means of an air-pump. It appeared to Papin that some other method might be found of effecting this end more expeditiously and with less labour. First he tried to produce the requisite vacuum by the explosion of a small quantity of gunpowder in the bottom of the cylinder, the momentary flame occasioned by which he thought would expel the air through a valve opening upwards in the piston, while the immediate fall of the valve, on the action of the flame being spent, would prevent its re-intrusion. But he never was able to effect a very complete vacuum by this method. He then, about the year 1690, bethought him of making use of steam for that purpose. This vapour, De Caus had long ago remarked, was recondensed and restored to the state of water by cold; but up to this time the attention of no person seems to have been awakened to the important advantage that might be taken of this one of its properties. Papin for the first time availed himself of it in his lifting machine, to produce the vacuum he wanted. Introducing a small quantity of water into the bottom of his cylinder, he heated it by a fire underneath, till it boiled and gave forth steam, which, by its powerful expansion, raised the piston from its original position in contact with the water, to a considerable height above it, even in opposition to the pressure of the atmosphere on its other side. This done, he then removed the fire, on which the steam again became condensed into water, and, occupying now about the seventeen hundredth part of its former dimensions, left a vacant space through which the piston was carried down by its own gravitation and the pressure of the atmosphere.

The machine thus proposed by Papin was abundantly defective in the subordinate parts of its mechanism, and, unimproved, could not have operated with much effect. But, imperfect as it was, it exemplified two new principles of the highest importance, neither of which appears to have been thought of, in the application of the power of steam, before his time. The first is the communication of the moving force of that agent to bodies upon which it cannot conveniently act directly, by means of the piston and its rod. The second is the deriving of the moving force desired, not from the expansion of steam, but from its other equally valuable property of condensibility by mere exposure to cold. Papin, however, it is curious enough, afterwards abandoned his piston and method of condensation, and reverted to the old plan of making the steam act directly by its expansive force upon the water to be raised. It is doubtful, however, whether he ever actually erected any working engine upon either of these constructions. Indeed, the improvement of the steam-engine could scarcely be said to have been the principal object of those experiments of his which, nevertheless, contributed so greatly to that result. It was, in fact, as we have seen, with the view of perfecting a machine contrived originally without any reference to the application of steam, that he was first induced to have recourse to the powers of that agent. The moving force with which he set out was the pressure of the atmosphere; and he employed steam merely as a means of enabling that other power to act. Even by such a seemingly subordinate application, however, of the new element, he happily discovered and bequeathed to his successors the secret of some of its most valuable capabilities.

We may here conveniently notice another ingenious contrivance, of essential service in the steam-engine, for which we are also indebted to Papin—we mean the safety-valve. This is merely a lid or stopper, closing an aperture in the boiler, and so loaded as to resist the expansive force of the steam up to a certain point, while at the same time, it must give way and allow free vent to the pent-up element, long before it can have acquired sufficient strength to burst the boiler. The safety-valve, however, was not introduced into the steam-engine either by Papin, or for some years after his time. It was employed by him only in the apparatus still known by the name of his *digester*, a contrivance for producing a very powerful heat in cookery and chemical preparations, by means of highly-concentrated steam.

We now come to the engine invented by Captain Savery in 1698. This gentleman, we are told, having one day drank a flask of Florence wine at a tavern, afterwards threw the empty flask upon the fire, when he was struck by perceiving that the small quantity of liquid still left in it very soon filled it with steam, under the influence of the heat. Taking it up again while thus full of vapour, he now plunged it, with the mouth downwards, into a basin of cold water, which happened to be on the table; by which means the steam being instantly concentrated, a vacuum was produced within the flask, into which the water immediately rushed up from the basin. According to another version of the story, it was the accidental circumstance of his immersing a heated tobacco-pipe into water, and perceiving the water immediately rush up through the tube, on the concentration by the cold of the warm and thin air, that first suggested to Savery the important use that might be made of steam, or any other gas expanded by heat, as a means of creating a vacuum. He did not, however, employ steam for this purpose in the same manner that Papin had done. Instead of a piston moving under the pressure of the atmosphere through the vacuum produced by the concentration of the steam, he availed himself of such a vacuum merely to permit the rise of the water into it from the well or mine below, exactly as in the common sucking-pump. Having thus raised the water to the level of the boiler, he afterwards allowed it to flow into another vessel, from whence he sent it to a greater height by the same method which had been many years before employed by the Marquis of Worcester,—namely, by making the expansive force of the steam act upon it directly, and so force it up in opposition to its own gravity and the resistance of the atmosphere.

Savery showed much ingenuity and practical skill in contriving means of facilitating and improving the working of the apparatus which he had devised upon these principles; and many of his engines were erected for supplying gentlemen's houses with water and other purposes, in different parts of the country. The machine also received many improvements after the death of the original inventor. It was considerably simplified, in particular, by Dr. Desaguliers, about the year 1718; and this gentleman also contrived a method of concentrating the steam by the injection of a small current of cold water into the receiver, instead of the old method employed by Savery, of dashing the water over the outside of the vessel, which cooled it to an unnecessary degree, and occasioned, therefore, a wasteful expenditure of fuel. It was Desaguliers who first introduced the safety-valve into the steam-engine, although Papin had previously suggested such an application of the contrivance. Engines upon Savery's principle have continued to be constructed, down to our own times; and, as they can be made at a comparatively small expense, they are found to answer very well in situations where water has to be raised only a short way. This engine is, in fact, merely a combination of the common sucking-pump (except that the requisite vacuum is produced by the condensation of steam, and without the aid of a piston) with the contrivance proposed by De Caus and the Marquis of Worcester for the application of the expansive force of steam; and, wherever the machine can be economically employed, the former part of it is that which operates with by far the most effect.

Not long after Savery had invented his engine, Thomas Newcomen, an ironmonger, and John Calley, a glazier, both of Dartmouth, in Devonshire, began also to direct their attention to the employment of steam as a mechanic power. Their first engine was constructed about the year 1711. This contrivance, which is commonly known by the name of Newcomen's engine, proceeded mainly upon the principle formerly adopted by Papin, but subsequently abandoned both by him and those who immediately followed him in the cultivation of this department of mechanics, of making the moving power of the machinery the weight of the atmosphere acting upon a piston, so as to carry it down through a vacuum created by the condensation of the steam. Newcomen's apparatus is, on this account, often distinguished by the name of the Atmospheric engine. Its inventors, however, instead of adopting Papin's clumsy method of cooling his steam by the removal of the fire, employed, in the first instance, the expedient of pouring cold water on the containing vessel, as Savery had done before them, though without being aware, it is said, of his prior claim to the improvement. They afterwards exchanged this for the still better method, already described as introduced by Desaguliers into Savery's engine, of injecting a stream of water into the cylinder, which is said to have been suggested to them by the accident of some water having found admission to the steam through a hole which happened to have worn itself in the piston. This engine of Newcomen, which in the course of a very few years after its invention was brought to as high a state of perfection as the principle seems to admit of, afforded the first important exemplification of the value of steam in mechanics. Savery's, the only other practical contrivance which had been proposed, had been found quite inadequate to the raising of water from any considerable depth, its principal power, as we have already remarked, lying, in fact, in the part of it which acted as a sucking-pump, and by which, as such, water could only be raised till its column was of equal weight with a column of the atmosphere of the same base. It was nearly useless, therefore, as an apparatus for pumping up water from mines; the grand object for which a moving force of extraordinary power was at this time in demand. But here Newcomen's engine proved of essential service. Many mines that had long remained unwrought were, immediately after its invention, again rendered accessible, and gradually excavated to great depths; while others were opened, and their treasures sought after with equal success, which but for its assistance could never have been attempted. It was applied also to various other important purposes.

Newcomen's engine, however, notwithstanding its usefulness, especially in cases where no other known power could be applied, was still in some respects a very defective contrivance, and by no means adapted to secure the complete command of the energies of steam. The great waste of fuel, in particular, which was still occasioned by the degree to which the cylinder was cooled after every stroke of the piston, from the cold water injected into it, rendered it scarcely any saving of expense to employ this engine in circumstances where animal power was available. Its whole force, too, the reader will observe, as a moving power, was limited to what could be obtained by atmospheric pressure alone, which, even could the vacuum under the piston have been rendered quite perfect, and all obstructions from friction annihilated, could only have amounted to about fifteen pounds for every square inch of the surface of the piston. The expansive force of steam was not, in fact, at all employed in this contrivance as a moving power; could the vacuum necessary to permit the descent of the piston have been as expeditiously and conveniently produced by any other agency, that of steam might have been dispensed with altogether. An air-pump, for instance, attached to the lower part of the cylinder, as originally proposed by Otto Guericke, might have rendered all the service which steam was here called upon to perform; and in that case, this element, with the fuel by which it was generated, might have been dispensed with, and the machine would not have been a steam-engine at all. This view of the matter may, in some degree, account for the complete neglect of steam as a moving power which so long prevailed after Newcomen's engine was brought into use, notwithstanding the proofs of its capabilities in that character which had been afforded by the attempts of the earlier speculators. It was now regarded simply as affording the easiest means of obtaining a ready vacuum, in consequence of its property of rapid condensation on the application of cold; its other property of extraordinary expansion, which had first attracted to it the attention of mechanicians, and presented in reality a much more obvious application of it as a mechanical agent, had been entirely neglected. The only improvements of the engine which were attempted or thought of were such as referred to what may be called its subordinate mechanism,—that is to say, the contrivances for facilitating the alternate supplies of the steam and the water on which its action depended; and after Mr. Beighton had, about the year 1718, made the machine itself shut and open the cocks by which these supplies were regulated, instead of having that service performed as at first by an attendant, there remained little more to be done even in this department. The steam might be applied with more ease and readiness, but not with any augmentation of effect; the power of the engine could be increased only by a more plentiful application of atmospheric pressure. It was with propriety, therefore, that Newcomen's invention was called, not a steam, but an atmospheric engine.

For half-a-century, accordingly, after the improvements introduced by Beighton, who may be considered as the perfector of this engine, no further progress worth mentioning was made in the application of steam as an agent in mechanics. The engine itself was more and more extensively employed, notwithstanding its defects; but no better method was proposed of calling into exercise the stupendous powers of the element, which, by means of only one of its remarkable properties, was here shown to be capable of rendering such valuable service. Our knowledge of what might be done by steam was in this state when the subject at last happily attracted the attention of Watt.

James Watt was born at Greenock, on the 19th of January, 1736. His father was a merchant, and also one of the magistrates of that town. He received the rudiments of his education in his native place; but his health being even then extremely delicate, as it continued to be to the end of his life, his attendance at school was not always very regular. He amply made up, however, for what he lost in this way by the diligence with which he pursued his studies at home, where without any assistance he succeeded at a very early age in making considerable proficiency in various branches of knowledge. Even at this time his favourite study is said to have been mechanical science, to a love of which he was probably in some degree led by the example of his grandfather and his uncle, both of whom had been teachers of the mathematics, and had left a considerable reputation for learning and ability in that department. Young Watt, however, was not indebted to any instructions of theirs for his own acquirements in science, the former having died two years before, and the latter the year after he was born. At the age of eighteen he was sent to London to be apprenticed to a maker of mathematical instruments; but in little more than a year the state of his health forced him to return to Scotland; and he never received any further instruction in his profession. A year or two after this, however, a visit which he paid to some relations in Glasgow suggested to him the plan of attempting to establish himself in that city in the line for which he had been educated. In 1757, accordingly, he removed thither, and was immediately appointed mathematical instrument maker to the College. In this situation he remained for some years, during which, notwithstanding almost constant ill health, he continued both to prosecute his profession, and to labour in the general cultivation of his mind, with extraordinary ardour and perseverance. Here also he enjoyed the friendship and intimacy of several distinguished persons who were then members of the University, especially of the celebrated Dr. Black, the discoverer of the principle of latent heat, and Mr. (afterwards Dr.) John Robison, so well known by his treatises on mechanical science, who was then a student and about the same age with himself. Honourable, however, as his present appointment was, and important as were many of the advantages to which it introduced him, he probably did not find it a very lucrative one; and therefore, in 1763, when about to marry, he removed from his apartments in the University to a house in the city, and entered upon the profession of a general engineer.

For this his genius and scientific attainments admirably qualified him. Accordingly, he soon acquired a high reputation, and was extensively employed in making surveys and estimates for canals, harbours, bridges, and other public works. His advice and assistance indeed were sought for in almost all the important improvements of this description which were now undertaken or proposed in his native country. But another pursuit, in which he had been for some time privately engaged, was destined ere long to withdraw him from this line of exertion, and to occupy his whole mind with an object still more worthy of its extraordinary powers.

While yet residing in the College his attention had been directed to the employment of steam as a mechanical agent by some speculations of his friend Mr. Robison, with regard to the practicability of applying it to the movement of wheel-carriages; and he had also himself made some experiments with Papin's digester, with the view of ascertaining its expansive force. He had not prosecuted the inquiry, however, so far as to have arrived at any determinate result, when, in the winter of 1763-64, a small model of Newcomen's engine was sent to him by the Professor of Natural Philosophy to be repaired, and fitted for exhibition in the class. The examination of this model set Watt upon thinking anew, and with more interest than ever, on the powers of steam.

The first thing that attracted his attention about the machine before him, the cylinder of which was only of two inches diameter, while the piston descended through six inches, was the insufficiency of the boiler, although proportionally a good deal larger than in the working engines, to supply the requisite quantity of steam for the creation of the vacuum. In order to remedy this defect he was obliged, in repairing the model, to diminish the column of water to be raised; in other words, to give the piston less to do, in compensation for its having to descend, not through a perfect vacuum, but in opposition to a considerable residue of undisplaced air. He also soon discovered the reason why in this instance the steam sent up from the boiler was not sufficient to fill the cylinder. In the first place, this containing vessel, being made, not of cast-iron, as in the larger engines, but of brass, abstracted more of the heat from the steam, and so weakened its expansion; and secondly, it exposed a much larger surface to the steam, in proportion to its capacity, than the cylinders of the larger engines did, and this operated still more strongly to produce the same effect. Led by the former of these considerations, he made some experiments in the first instance with the view of discovering some other material whereof to form the cylinder of the engine which should be less objectionable than either brass or cast-iron; and he proposed to substitute wood, soaked in oil, and baked dry. But his speculations soon took a much wider scope; and, struck with the radical imperfections of the atmospheric engine, he began to turn in his mind the possibility of employing steam in mechanics, in some new manner which should enable it to operate with much more powerful effect. This idea having got possession of him, he engaged in an extensive course of experiments, for the purpose of ascertaining as many facts as possible with regard to the properties of steam; and the pains he took in this investigation were rewarded with several valuable discoveries. The rapidity with which water evaporates, he found, for instance, depended simply upon the quantity of heat which was made to enter it; and this again on the extent of the surface exposed to the fire. He also ascertained the quantity of coals necessary for the evaporation of any given quantity of water, the heat at which water boils under various pressures, and many other particulars of a similar kind which had never before been accurately determined.

Thus prepared by a complete knowledge of the properties of the agent with which he had to work, he next proceeded to take into consideration, with a view to their amendment, what he deemed the two grand defects of Newcomen's engine. The first of these was the necessity arising from the method employed to concentrate the steam, of cooling the cylinder, before every stroke of the piston, by the water injected into it. On this account, a much more powerful application of heat than would otherwise have been requisite was demanded for the purpose of again heating that vessel when it was to be refilled with steam. In fact, Watt ascertained that there was thus occasioned, in the feeding of the machine, a waste of not less than three-fourths of the whole fuel employed. If the cylinder, instead of being thus cooled for every stroke of the piston, could be kept permanently hot, a fourth part of the heat which had been hitherto applied would be found to be sufficient to produce steam enough to fill it. How, then, was this desideratum to be attained? De Caus had proposed to effect the condensation of the steam by actually removing the furnace from under the boiler before every stroke of the piston; but this, in a working engine, evidently would have been found quite impracticable. Savery, the first who really constructed a working engine, and whose arrangements, as we have already remarked, all showed a very superior ingenuity, employed the method of throwing cold water over the outside of the vessel containing his steam—a perfectly manageable process, but at the same time a very wasteful one; inasmuch as, every time it was repeated, it cooled not only the steam, but the vessel also, which, therefore, had again to be heated by a large expenditure of fuel before the steam could be reproduced. Newcomen's method of injecting the water into the cylinder was a considerable improvement on this; but it was still objectionable on the same ground, though not to the same degree; it still cooled not only the steam, on which it was desired to produce that effect, but also the cylinder itself, which, as the vessel in which more steam was to be immediately manufactured, it was so important to keep hot. It was also a very serious objection to this last-mentioned plan, that the injected water itself, from the heat of the place into which it was thrown, was very apt to be partly converted into steam; and the more cold water was used, the more considerable did this creation of new steam become. In fact, in the best of Newcomen's engines, the perfection of the vacuum was so greatly impaired from this cause, that the resistance experienced by the piston in its descent was found to amount to about a fourth part of the whole atmospheric pressure by which it was carried down, or, in other words, the working power of the machine was thereby diminished one-fourth.

After reflecting for some time upon all this, it at last occurred to Watt to consider whether it might not be possible, instead of continuing to condense the steam in the cylinder, to contrive a method of drawing it off, to undergo that operation in some other vessel. This fortunate idea having presented itself to his thoughts, it was not very long before his ingenuity also suggested to him the means of realising it. In the course of one or two days, according to his own account, he had all the necessary apparatus arranged in his mind. The plan which he devised, indeed, was an extremely simple one, and on that account the more beautiful. He proposed to establish a communication by an open pipe between the cylinder and another vessel, the consequence of which evidently would be, that when the steam was admitted into the former, it would flow into the latter so as to fill it also. If, then, the portion in this latter vessel only should be subjected to a condensing process, by being brought into contact with cold water, or any other convenient means, what would follow? Why, a vacuum would be produced here—into that, as a vent, more steam would immediately rush from the cylinder—that likewise would be condensed—and so the process would go on till all the steam had left the cylinder, and a perfect vacuum had been effected in that vessel, without so much as a drop of cold water having touched or entered it. The separate vessel alone, or the condenser, as Watt called it, would be cooled by the water used to condense the steam—and that, instead of being an evil, manifestly tended to promote and quicken the condensation. When Watt reduced these views to the test of experiment, he found the result to answer his most sanguine expectations. The cylinder, although emptied of its steam for every stroke of the piston as before, was now constantly kept at the same temperature with the steam (or 212° Fahrenheit); and the consequence was, that one-fourth of the fuel formerly required sufficed to feed the engine. But besides this most important saving in the expense of maintaining the engine, its power was greatly increased by the more perfect vacuum produced by the new construction, in which the condensing water, being no longer admitted within the cylinder, could not, as before, create new steam there while displacing the old. The first method which Watt adopted of cooling the steam in the condenser was to keep that vessel surrounded by cold water—considering it as an objection to the admission of the water into its interior, that it might be difficult in that case to convey it away as fast as it would accumulate. But he found that the condensation was not effected in this manner with so much rapidity as was desirable. It was necessary for him, too, at any rate to employ a pump attached to the condenser, in order to draw off both the small quantity of water deposited by the cooled steam, and the air unavoidably introduced by the same element—either of which, if allowed to accumulate, would have impaired the perfect vacuum necessary to attract the steam from the cylinder. He therefore determined eventually to admit also the additional quantity of water required for the business of condensation, and merely to employ a larger and more powerful pump to carry off the whole.

Such, then, was the remedy by which the genius of this great inventor effectually cured the first and most serious defect of the old apparatus. In carrying his ideas into execution, he encountered, as was to be expected, many difficulties arising principally from the impossibility of realising theoretical perfection of structure with such materials as human art is obliged to work with; but his ingenuity and perseverance overcame every obstacle. One of the things which cost him the greatest trouble was, how to fit the piston so exactly to the cylinder as without affecting the freedom of its motion, to prevent the passage of the air between the two. In the old engine this end had been attained by covering the piston with a small quantity of water, the dripping down of which into the space below, where it merely mixed with the stream introduced to effect the condensation, was of little or no consequence. But in the new construction, the superiority of which consisted in keeping this receptacle for the steam always both hot and dry, such an effusion of moisture, although only in very small quantities, would have occasioned material inconvenience. The air alone besides, which in the old engine followed the piston in its descent, acted with considerable effect in cooling the lower part of the cylinder. His attempts to overcome this difficulty, while they succeeded in that object, conducted Watt also to another improvement, which effected the complete removal of what we have called the second radical imperfection of Newcomen's engine—namely, its non-employment, for a moving power, of the expansive force of the steam. The effectual way, it occurred to him, of preventing any air from escaping into the part of the cylinder below the piston, would be to dispense with the use of that element above the piston, and to substitute there likewise the same contrivance as below, of alternate steam and a vacuum. This was, of course, to be accomplished by merely opening communications from the upper part of the cylinder to the boiler on the one hand, and the condenser on the other, and forming it at the same time into an air-tight chamber, by means of a cover, with only a hole in it to admit the rod or shank of the piston, which might, besides, without impeding its freedom of action, be padded with hemp, the more completely to exclude the air. It was so contrived accordingly, by a proper arrangement of the cocks and the machinery connected with them, that, while there was a vacuum in one end of the cylinder, there should be an admission of steam into the other; and the steam so admitted now served, not only, by its susceptibility of sudden condensation, to create the vacuum, but also, by its expansive force, to impel the piston. Steam, in fact, was now restored to be, what it had been in the early attempts to use it as a mechanical agent, the moving power of the engine; but its efficiency in this capacity was for the first time both taken full advantage of, by means of contrivances properly arranged for that end, and combined with, and aided by, its other equally valuable property which had alone been called into action in the more recent machines.

These were the great improvements which Watt introduced in what may be called the principle of the steam-engine, or, in other words, in the manner of using and applying the steam. They constitute, therefore, the grounds of his claim to be regarded as the true author of the conquest that has at last been obtained by man over this powerful element. But original and comprehensive as were the views out of which these fundamental inventions arose, the exquisite and inexhaustible ingenuity which the engine, as finally perfected by him, displays in every part of its subordinate mechanism, is calculated to strike us, perhaps, with scarcely less admiration. It forms, undoubtedly, the best exemplification that has ever been afforded of the number and diversity of services which a piece of machinery may be made to render to itself by means solely of the various application of its first moving power, when that has once been called into action. Of these contrivances, however, we can only notice one or two, by way of specimen. Perhaps the most singular is that called the *governor*. This consists of an upright spindle, which is kept constantly turning, by being connected with a certain part of the machinery, and from which two balls are suspended in opposite directions by rods, attached by joints, somewhat in the manner of the legs of a pair of tongs. As long as the motion of the engine is uniform, that of the spindle is so likewise, and the balls continue steadily revolving at the same distance from each other. But as soon as any alteration in the action of the piston takes place, the balls, if it has become more rapid, fly farther apart under the influence of the increased centrifugal force which actuates them—or approach nearer to each other in the opposite circumstances. This alone would have served to indicate the state of matters to the eye; but Watt was not to be so satisfied. He connected the rods with a valve in the tube by which the steam is admitted to the cylinder from the boiler, in such a way, that as they retreat from each other, they gradually narrow the opening which is so guarded, or enlarge it as they tend to collapse; thus diminishing the supply of steam when the engine is going too fast, and, when it is not going fast enough, enabling it to regain its proper speed by allowing it an increase of aliment. Again, the constant supply of a sufficiency of water to the boiler is secured by an equally simple provision—namely, by a *float* resting on the surface of the water, which, as soon as it is carried down by the consumption of the water to a certain point, opens a valve and admits more. And so on, through all the different parts of the apparatus, the various wonders of which cannot be better summed up than in the forcible and graphic language of a recent writer :—" In the present perfect state of the engine it appears a thing almost endowed with intelligence. It regulates with perfect accuracy and uniformity the *number of its strokes* in a given time, *counting* or *recording* them, moreover, to tell how much work it has done, as a clock records the beats of its pendulum;—it regulates the *quantity of steam* admitted to work;—the *briskness of the fire;*—the *supply of water* to the boiler;—the *supply of coals* to the fire;—it *opens and shuts its valves* with absolute precision as to time and manner;—it *oils its joints;*—it *takes out any air* which may accidentally enter into parts which should be vacuous; and when anything goes wrong which it cannot of itself rectify, it *warns its attendants* by ringing a bell; yet with all these talents and qualities, and even when exerting the power of six hundred horses, it is obedient to the hand of a child;—its aliment is coal, wood, charcoal, or other combustible,—it consumes none while idle,—it never tires, and wants no sleep;—it is not subject to malady when originally well made, and only refuses to work when worn out with age; it is equally active in all climates, and will do work of any kind;—it is a water-pump, a miner, a sailor, a cotton-spinner, a weaver, a blacksmith, a miller, &c. &c. ; and a small engine, in the character of a *steam pony*, may be seen dragging after it on a railroad a hundred tons of merchandise, or a regiment of soldiers, with greater speed than that of our fleetest coaches. It is the king of machines, and a permanent realisation of the *Genii* of Eastern fable, whose supernatural powers were occasionally at the command of man."

In addition to those difficulties which his unrivalled mechanical ingenuity enabled him to surmount, Watt, notwithstanding the merit of his inventions, had to contend for some time with others of a different nature, in his attempts to reduce them to practice. He had no pecuniary resources of his own, and was at first without any friend willing to run the risk of the outlay necessary for an experiment on a sufficiently large scale. At last he applied to Dr. Roebuck, an ingenious and spirited speculator, who had just established the Carron iron-works, not far from Glasgow, and held also at this time a lease of the extensive coal-works at Kinneal, the property of the Duke of Hamilton. Dr. Roebuck agreed to advance the requisite funds on having two-thirds of the profits made over to him; and upon this Mr. Watt took out his first patent in the beginning of the year 1769. An engine with a cylinder of eighteen inches diameter was soon after erected at Kinneal; and although, as a first experiment, it was necessarily in some respects of defective construction, its working completely demonstrated the great value of Watt's improvements. But Dr. Roebuck, whose undertakings were very numerous and various, in no long time after forming this connection, found himself involved in such pecuniary difficulties, as to put it out of his power to make any further advances in prosecution of its object. On this Watt employed himself for some years almost entirely to the ordinary work of his profession as a civil engineer; but at last, about the year 1774, when all hopes of any further assistance from Dr. Roebuck were at an end, he resolved to close with a proposal which had been made to him through his friend Dr. Small, of Birmingham, that he should remove to that town, and enter into partnership with the eminent hardware manufacturer, Mr. Boulton, whose extensive establishments at Soho had already become famous over Europe, and procured for England an unrivalled reputation for the arts there carried on. Accordingly, an arrangement having been made with Dr. Roebuck, by which his share of the patent was transferred to Mr. Boulton, the firm of Boulton & Watt commenced the business of making steam-engines in the year 1775.

Mr. Watt now obtained from Parliament an extension of his patent for twenty-five years from this date, in consideration of the acknowledged national importance of his inventions. The first thing which he and his partner did was to erect an engine at Soho, which they invited all persons interested in such machines to inspect. They then proposed to erect similar engines wherever required, on the very liberal principle of receiving as payment for each, only one-third of the saving in fuel which it should effect, as compared with one of the old construction. As this saving, however, had been found to amount in the whole to fully three-fourths of all the fuel that had been wont to be employed, the revenue thus accruing to the patentees became very great after their engines were extensively adopted This they very soon were, especially in Cornwall, where the numerous mines afforded a vast field for the employment of the new power, partly in continuing or commencing works which only an economised expenditure could make profitable, and often also in labours which the old engine was altogether inadequate to attempt.

But the draining of mines was only one of many applications of the steam power now at his command which Watt contemplated, and in course of time accomplished. During the whole twenty-five years, indeed, over which his renewed patent extended, the perfecting of his invention was his chief occupation; and, notwithstanding a delicate state of health, and the depressing affliction of severe headaches to which he was extremely subject, he continued throughout this period to persevere with unwearied diligence in adding new improvements to the mechanism of the engine, and devising the means of applying it to new purposes of usefulness. He devoted, in particular, the exertions of many years to the contriving of the best methods of making the action of the piston communicate a rotatory motion in various circumstances, and between the years 1781 and 1785 he took out four different patents for inventions having this object in view. In the midst of these scientific labours, too, his attention was much distracted by attempts which were made in several quarters to pirate his improvements, and the consequent necessity of defending his rights in a series of actions, which, notwithstanding successive verdicts in his favour, did not terminate till the year 1799, when the validity of his claims was finally confirmed by the unanimous decision of the Judges of the Court of King's Bench.

Watt's inexhaustible ingenuity displayed itself in various other contrivances beside those which make part of his steam-engine. An apparatus for copying letters and other writings, now in extensive use; a method of heating houses by steam; a new composition, for the purposes of sculpture, having the transparency and nearly the hardness of marble; a machine for multiplying copies of busts and other performances in carving or statuary,—are enumerated among his minor inventions. But it is his steam-engine that forms the great monument of his genius, and that has conferred upon his name its imperishable renown. This invention has already gone far to revolutionise the whole domain of human industry; and almost every year is adding to its power and its conquests. In our manufactures, our arts, our commerce, our social accommodations, it is constantly achieving what, little more than half-a-century ago, would have been counted miracles and impossibilities. "The trunk of an elephant, it has been finely and truly said, that can pick up a pin, or rend an oak, is as nothing to it. It can engrave a seal, and crash masses of obdurate metal like wax before it,—draw out, without breaking, a thread as fine as gossamer,—and lift a ship of war like a bauble in the air. It can embroider muslin and forge anchors; cut steel into ribbands, and impel loaded vessels against the fury of the winds and waves." And another application of it, which was made but a short time afterwards, was destined to be productive of still greater changes on the condition of society than have resulted from any of its previous achievements. It had been employed, several years previously, at some of our collieries, in the propelling of heavily-loaded carriages over railways; but the great experiment of the Liverpool and Manchester Railway (opened in 1830) had, for the first time, practically demonstrated with what hitherto almost undreamt-of *rapidity* travelling by land may hereafter be carried on through the aid of steam. Coaches, under the impetus communicated by this, the most potent, and at the same time the most perfectly controllable of all our mechanical agencies, were destined to be drawn forward at the flying speed of thirty and thirty-five miles an-hour. When so much has been done already, it would be rash to conclude that even this is to be our ultimate limit of attainment. In navigation, the resistance of the water, which increases rapidly as the force opposed to it increases, very soon sets bounds to the rate at which even the power of steam can impel a vessel forward. But, on land, the thin medium of the air presents no such insurmountable obstacle to a force making its way through it; and a rapidity of movement may perhaps be eventually attained here, which is to us even as yet inconceivable. But even when the rate of land travelling already shown to be quite practicable shall have become universal in what a new state of society shall we find ourselves! When we are able to travel a hundred miles in any direction in six or eight hours, into what comparative neighbourhood will the remotest extremes even of a large country be brought, and how little shall we think of what we now call distance! A nation will then be, indeed, a community; and all the benefits of the highest civilisation, instead of being confined to one central spot, will be diffused equally over the land, like the light of heaven. This improvement, in short, when fully consummated, confers upon man nearly as much new power and new enjoyment as if he were actually endowed with wings.

It is gratifying to reflect that even while he was yet alive, Watt received from the voice of the most illustrious of his contemporaries the honours due to his genius. In 1785 he was elected a Fellow of the Royal Society; the degree of Doctor of Laws was conferred upon him by the University of Glasgow in 1806; and in 1808 he was elected a member of the French Institute. He died on the 25th of August, 1819, in the 84th year of his age.

We cannot better conclude our sketch of the life of this great inventor than by the following extract from the character that has been drawn of him by the eloquent writer (Lord Jeffrey) whom we have already quoted :—

" Independently of his great attainments in mechanics, Mr Watt was an extraordinary, and in many respects a wonderful man. Perhaps no individual in his age possessed so much and such varied and exact information,—had read so much, or remembered what he had read so accurately and well. He had infinite quickness of apprehension, a prodigious memory, and a certain rectifying and methodising power of understanding, which extracted something precious out of all that was presented to it. His stores of miscellaneous knowledge were immense, and yet less astonishing than the command he had at all times over them. It seemed as if every subject that was casually started in conversation had been that which he had been last occupied in studying and exhausting; such was the copiousness, the precision, and the admirable clearness of the information which he poured out upon it without effort or hesitation. Nor was this promptitude and compass of knowledge confined in any degree to the studies connected with his ordinary pursuits. That he should have been minutely and extensively skilled in chemistry and the arts, and in most of the branches of physical science, might perhaps have been conjectured; but it could not have been inferred from his usual occupations, and probably is not generally known, that he was curiously learned in many branches of antiquity, metaphysics, medicine, and etymology, and perfectly at home in all the details of architecture, music, and law. He was well acquainted, too, with most of the modern languages, and familiar with their most recent literature. Nor was it at all extraordinary to hear the great mechanician and engineer detailing and expounding, for hours together, the metaphysical theories of the German logicians, or criticising the measures or the matter of the German poetry.

"His astonishing memory was aided, no doubt, in a great measure, by a still higher and rarer faculty—by his power of digesting and arranging in its proper place all the information he received, and of casting aside and rejecting, as it were instinctively, whatever was worthless or immaterial. Every conception that was suggested to his mind seemed instantly to take its place among its other rich furniture, and to be condensed into the smallest and most convenient form. He never appeared, therefore, to be at all encumbered or perplexed with the *verbiage* of the dull books he perused, or the idle talk to which he listened; but to have at once extracted, by a kind of intellectual alchemy, all that was worthy of attention, and to have reduced it for his own use to its true value and to its simplest form.

And thus it often happened, that a great deal more was learned from his brief and vigorous account of the theories and arguments of tedious writers, than an ordinary student could ever have derived from the most faithful study of the originals, and that errors and absurdities became manifest from the mere clearness and plainness of his statement of them, which might have deluded and perplexed most of his hearers without that invaluable assistance."

The monumental inscription written by Lord Brougham for the statue of Watt in Westminster Abbey, is as follows:—

> Not to perpetuate a Name
> Which must endure while the peaceful Arts flourish
> But to show
> That mankind have learned to honour those
> Who best deserve their gratitude
> The King
> His Ministers and many of the Nobles
> And Commons of the Realm
> Raised this monument to
> JAMES WATT
> Who directing the force of an original genius
> Early exercised in philosophic research
> To the improvement of the
> Steam-engine
> Enlarged the resources of his country
> Increased the power of man
> And rose to an eminent place
> Among the most illustrious followers of Science
> And the real benefactors of the world.

SIR HUMPHREY DAVY.

SIR HUMPHREY DAVY was born in 1778, at Penzance, in Cornwall. His father followed the profession of a carver in wood in that town, where many of his performances are still to be seen in the houses of the inhabitants. All that we are told of Davy's school education is, that he was taught the rudiments of classical learning at a seminary in Truro. He was then placed by his father with an apothecary and surgeon in his native place. But, instead of attending to his profession, he spent his time either in rambling about the country or experimenting in his master's garret, sometimes to the no small danger of the whole establishment; and the doctor and he at last agreed to part. About his fifteenth year he was placed as pupil with another surgeon residing in Penzance; but it does not appear that his second master had much more success than his first in attempting to give him a liking for the medical profession. The future philosopher, however, had already begun to devote himself, of his own accord, to those sciences in which he afterwards so greatly distinguished himself; and proceeding upon a plan of study which he had laid down for himself, he had, by the time he was eighteen, obtained a thorough knowledge of the rudiments of natural philosophy and chemistry, as well as made some proficiency in botany, anatomy, and geometry. The subject of metaphysics, it is stated, was also embraced in his reading at this period.

But chemistry was the science to which, of all others, he gave himself with the greatest ardour; and, even in this early stage of his researches, he seems to have looked forward to fame from his labours in this department. The writer of the memoir of Sir Humphrey to which we are indebted for these particulars, quotes an exclamation which broke from him one day in after-life, when contemplating, along with a friend, a picture of one of the mines of his native district, which shows what were the visions of his solitary rambles. "How often, when a boy," said he, "have I wandered about those rocks in search after new minerals, and, when tired, sat down upon those crags, and exercised my fancy in anticipations of future renown!" The peculiar features of this part of the country doubtless contributed not a little to give his genius the direction it took. The mineral riches concealed under the soil formed alone a world of curious investigation. The rocky coast presented a geological study of inexhaustible interest. Even the various productions cast ashore by the sea were continually affording new materials of examination to his inquisitive and reflecting mind. The first original experiment, it is related, in which he engaged, had for its object to ascertain the nature of the air contained in the bladders of sea-weed. At this time he had no other laboratory than what he contrived to furnish for himself, by the assistance of his master's phials and gallipots, the pots and pans used in the kitchen, and such other utensils as accident threw in his way. These he converted, with great ingenuity, to his own purposes. On one occasion, however, he accounted himself particularly fortunate in a prize which he made. This was a case of surgical instruments with which he was presented by the surgeon of a French vessel that had been wrecked on the coast, to whom he had done some kind offices. Examining his treasure with eagerness, Davy soon perceived the valuable aid he might derive in his philosophical experiments from some of the articles; and one of the principal of them was, in no long time, converted into a tolerable air-pump. The proper use of the instruments was, of course, as little thought of by their new possessor as that of his master's gallipots was wont to be when he had got them up to his garret. Davy's subsequent success as an experimentalist, it is well remarked by the writer to whom we have referred above, was probably owing, in no small degree, to the necessity he was placed under in his earliest researches of exercising his skill and ingenuity in this fashion. "Had he," proceeds his biographer, "in the commencement of his career, been furnished with all those appliances which he enjoyed at a later period, it is more than probable that he might never have acquired that wonderful tact of manipulation, that ability of suggesting expedients, and of contriving apparatus so as to meet and surmount the difficulties which must constantly arise during the progress of the philosopher through the unbeaten tracks and unexplored regions of science. In this art Davy certainly stands unrivalled; and, like his prototype, Scheele, he was unquestionably indebted for his address to the circumstances which have been alluded to: there never, perhaps, was a more striking exemplification of the adage, that necessity is the parent of invention."

A curious catalogue might be made of the shifts to which ingenious students in different departments of art have resorted, when, like Davy, they have wanted the proper instruments for carrying on their inquiries or experiments. His is not the first case in which the stores of an apothecary's shop are recorded to have fed the enthusiasm and materially assisted the labours of the young cultivator of natural science. The German chemist, Scheele, who has just been mentioned, and whose name ranks in his own department with the greatest of his time, was, as well as Davy, apprenticed in early life to an apothecary. While living in his master's house he used secretly to prosecute the study of his favourite science by employing often half the night in reading the works that treated of it, or making experiments with instruments fabricated, as Davy's were, by himself, and out of equally simple materials. Like the young British philosopher, too, Scheele is recorded to have sometimes alarmed the whole household by his detonations;—an incident which always brought down upon him the severe anger of his master, and heavy menaces intended to deter him from ever again applying himself to such dangerous studies, which, however, he did not long regard. It was at an apothecary's house, as has been noticed in a former page, that Boyle and his Oxford friends first held their scientific meetings, induced, as we are expressly told, by the opportunity they would thus have of obtaining drugs wherewith to make their experiments. Newton lodged with an apothecary, while at school, in the town of Grantham; and as, even at that early age, he is known to have been ardently devoted to scientific contrivances and experiments, and to have been in the habit of converting all sorts of articles into auxiliaries in his favourite pursuits, it is not probable that the various strange preparations which filled the shelves and boxes of his landlord's shop would escape his curious examination. Although Newton's glory chiefly depends upon his discoveries in abstract and mechanical science, some of his speculations, and especially some of his writings on the subjects of light and colour, show that the internal constitution of matter and its chemical properties had also much occupied his thoughts. Thus, too, in other departments, genius has found its sufficient materials and instruments in the humblest and most common articles, and the simplest contrivances. Fergusson observed the places of the stars by means of a thread with a few beads strung on it, and Tycho Brahe did the same thing with a pair of compasses. The self-taught American philosopher, Rittenhouse, being, when a young man, employed as an agricultural labourer, used to draw geometrical diagrams on his plough, and study them as he turned up the furrow. Pascal, when a mere boy, made himself master of many of the elementary propositions of geometry, without the assistance of any master, by tracing the figures on the floor of his room with a bit of coal. This, or a stick burned at the end, has often been the young painter's first pencil, while the smoothest and whitest wall he could find supplied the place of a canvas. Such, for example, were the commencing essays of the early Tuscan artist, Andrea del Castagno, who employed his leisure in this manner when he was a little boy tending cattle, till his performances at last attracted the notice of one of the Medici family, who placed him under a proper master. The famous Salvator Rosa first displayed his genius for design in the same manner. To these instances may be added that of the English musical composer Mr. John Davy, who is said, when only six years old, to have begun the study and practice of his art by imitating the chimes of a neighbouring church with eight horse-shoes, which he suspended by strings from the ceiling of a room in such a manner as to form an octave.

But to return to the subject of our notice. Davy first pursued his chemical studies without teacher or guide, in the manner that has been described, and aided only by the scantiest and rudest apparatus. When still a lad, however, he was fortunate in making the acquaintance of Mr. Gregory Watt, the son of the celebrated James Watt. This gentleman, having come to reside at Penzance for the recovery of his health, lodged with Mrs. Davy, and soon discovered the talent of her son. The scientific knowledge of Mr. Watt gave an accurate direction to the studies of the young chemist, and excited him to a systematic perseverance in his favourite pursuit. Chance attracted to him the notice of Mr. Davies Giddy (afterwards Mr. Gilbert, and President of the Royal Society), which the discovery of his merits soon improved into patronage and friendship. The boy, we are told, was leaning on the gate of his father's house when Mr. Gilbert passed accompanied by some friends, one of whom remarked that there was young Davy, who was so much attached to chemistry. The mention of chemistry immediately fixed Mr. Gilbert's attention; he entered into conversation with the young man, and, speedily becoming convinced of his extraordinary talents and acquirements, offered him the use of his library, and whatever other assistance he might require for the pursuit of his studies. Mr. Gilbert and Mr. Watt soon after this introduced Davy to the celebrated Dr. Beddoes, who had just established at Bristol what he called his Pneumatic Institution, for investigating the medical properties of the different gases. Davy, who was now in his nineteenth year, had for some time been thinking of proceeding to Edinburgh in order to pursue a regular course of medical education; but Dr. Beddoes, who had been greatly struck by different proofs he had given of his talents, and especially by an essay in which he expounded an original theory of light and heat, having offered him the superintendence of his new institution, he at once closed with that proposal.

The young philosopher was now fairly entered on his proper path, and, from this period, we may consider him as having escaped from the disadvantages of his early lot. But it was while yet poor and unknown that he had made those acquirements which both obtained for him the notice of his present patrons, and fitted him for the situation in which they placed him. His having attracted the attention of Mr. Gilbert, as he stood at his father's gate, may be called a fortunate incident; but it was one that never would have happened had it not been for the proficiency he had already made in science by his own endeavours. Chance may be said to have offered this opportunity of emerging from obscurity; but, had he not previously laboured in the cultivation of his mind as he had done, it would to him have been no opportunity at all.

The experiments conducted by Davy, and under his direction, at the Bristol Institution, were soon rewarded by important results; and of these, Davy, when he had just completed his twenty-first year, published an account, under the title of " Researches, Chemical and Philosophical, chiefly concerning Nitrous Oxide, and its respiration." In this publication the singularly intoxicating effects produced by the breathing of nitrous oxide were first announced, and it excited a considerable sensation in the scientific world, and at once made Davy generally known as a most ingenious and philosophic experimentalist. He was, in consequence, soon after its appearance, invited to fill the chemical chair of the Royal Institution, then newly established. When he commenced his lectures here, he was scarcely twenty-two years of age; but never was success in such an attempt more decided and brilliant. He soon saw his lecture-rooms crowded day after day by all that was most distinguished in the rank and intellect of the metropolis; and his striking and beautiful elucidations of every subject that came under his review, riveted, often even to breathlessness, the attention of his splendid auditory. The year after his appointment to this situation he was elected also Professor of Chemistry to the Board of Agriculture; and he greatly distinguished himself by the lectures which, for ten successive sessions, he delivered in this character. They were published in 1813 at the request of the Board. In 1803, when only in his twenty-fifth year, Davy was elected a Fellow of the Royal Society, and his contributions to the transactions from this time till his death were frequent, and of the highest value. In 1806 he was chosen to deliver the Bakerian Lecture before the Society; and he performed the same task for several successive years. Many of his most brilliant discoveries were announced in these discourses. In 1812 he received the honour of knighthood from the Prince Regent, being the first person on whom his Royal Highness conferred that dignity; and two days after he married a lady who brought him a considerable fortune. Next year he was elected a corresponding member of the French Institute. He was created a baronet in 1818. In 1820 he was chosen a foreign associate of the Royal Academy of Sciences at Paris on the death of the illustrious Watt. He had been for some time secretary to the Royal Society; and in 1820, on the death of Sir Joseph Banks, he was, by a unanimous vote, raised to the presidency of that learned body—an office which he held till he was obliged to retire, from ill health, in 1827, when his friend and first patron, Mr. Gilbert, was chosen to succeed him. Little, we may suppose, did either of the two anticipate, when they first met, thirty years before, at the gate of Davy's father's house, that they would thus stand successively, and in this order, at the head of the most distinguished scientific association in England.

It is impossible for us in this place to attempt anything more than the most general sketch of Sir Humphrey Davy's numerous and most important discoveries in chemical science. Even his earliest publication, the title of which we have already transcribed, was regarded as, for the first time, introducing light and order into an interesting department of the science,—the theory of the various combinations of oxygen and nitrogen, the two gases which, mixed together in certain proportions, form our common atmospheric air, but in other proportions produce compounds of an altogether dissimilar character. The first memoir by Davy which was read before the Royal Society was presented by him in 1801, before he was a member. It announced a new theory, which is now generally received, of the galvanic influence, or the extraordinary effect produced by two metals in contact with each other, when applied to the muscle even of a dead animal, which the Italian professor, Galvani, had some years before accidentally discovered. It was supposed, both by Galvani and his countryman Volta, who also distinguished himself in the investigation of this curious subject, that the effect in question was an electrical phenomenon—whence galvanism used to be called animal electricity; but Davy showed, by many ingenious experiments, that, in order to produce it, the metals in fact underwent certain chemical changes. Indeed, he proved that the effect followed when only one metal was employed, provided the requisite chemical change was by any means brought about on it—as, for example, by the interposition between two plates of it, of a fluid calculated to act upon its surface in a certain manner. In his Bakerian Lecture for 1806, he carried the examination of this subject to a much greater length, and astonished the scientific world by the announcement of a multitude of the most extraordinary results, from the application of the galvanic energy to the composition and decomposition of various chemical substances. From these experiments he arrived at the conclusion, that the power called chemical affinity was in truth, identical with that of electricity. Hence the creation of a new science, now commonly known by the name of Electro-Chemistry, being that which regards the supposed action of electricity in the production of chemical changes. The discourse in which these discoveries were unfolded was crowned by the French Institute with their first prize, by a decision which reflects immortal honour upon that illustrious body; who thus forgot not only all feelings of national jealousy, but even the peculiar and extraordinary hostility produced by the war which then raged between the two countries, in their admiration of genius and their zeal for the interests of philosophy.

But the results which this great chemist had already obtained only formed, in his hands, the source of new discoveries. In the interesting and extraordinary nature of its announcements, the Bakerian Lecture of 1807 was as splendid a production as that of the former year. There are certain substances, as the reader is aware, known in chemistry by the name of alkalis, of which potash and soda are the principal. These substances, chemists had hitherto in vain exhausted their ingenuity, and the resources of their art, in endeavouring to decompose. The only substance possessing alkaline properties, the composition of which had been ascertained, was ammonia, which is a gas, and is therefore called volatile alkali; and this having been found to be a compound of certain proportions of hydrogen and nitrogen, an opinion generally prevailed that hydrogen would be found to be also a chief ingredient of the *fixed* alkalis. Davy determined, if possible, to ascertain this point, and engaged in the investigation with great hopes of success, from the surpassing powers of decomposition which he had found to belong to his new agent, the galvanic influence. The manner in which he pursued this object is one of the most beautiful specimens of scientific investigation on record. One of the most important of the laws of galvanic decomposition, which he had previously discovered, was that, when any substance was subjected to this species of action, its oxygen (an ingredient which nearly all substances contain) was developed at what is called the positive end or pole of the current of electricity, while, whenever any hydrogen or inflammable matter was present, it uniformly appeared at the opposite or negative pole. Proceeding upon this principle, therefore, Davy set to work with a fixed alkali; and at first submitted it dissolved in water to the galvanic action. The result, however, was, that the water alone was decomposed, nothing being disengaged by the experiment but oxygen and hydrogen, the ingredients of that fluid, which passed off as usual, the former at the positive, the latter at the negative pole. In his subsequent experiments, therefore, Davy proceeded without water, employing potash in a state of fusion; and, having guarded the process from every other disturbing cause that presented itself, by a variety of ingenious arrangements, he had at last the satisfaction of seeing the oxygen gas developed, as before, at the positively electrified surface of the alkali, while at the same time, on the other side, small globules of matter were disengaged, having all the appearances of a metal. The long-agitated question was now determined; the base of the fixed alkalis was clearly metallic.

To ascertain the qualities of the metallic residue which he had thus obtained from the potash, was Davy's next object. From its great attraction for oxygen, it almost immediately, when exposed to the atmosphere, became an alkali again, by uniting with that ingredient; and at first it seemed on this account hardly possible to obtain a sufficient quantity of it for examination. But at last Davy thought of pouring over it a thin coating of the mineral fluid called naphtha, which both preserved it from communication with the air, and, being transparent, allowed it to be examined.

We have thus rapidly sketched the course of these brilliant and successful experiments, because they form a most interesting and instructive exemplification of the manner in which knowledge is pursued, and the secrets of nature extorted from her by well-directed interrogation. The business of philosophic experiment, it may be well to observe, is not a mere random expenditure of tests and applications. The true disciple of the inductive philosophy, on the contrary, has always in his contemplation, while conducting his experiments, an idea or end which he aims at realising, and which, in fact, directs him to every experiment to which he resorts. Thus, in the present instance, the idea in Davy's mind was, that the alkali was compounded of two ingredients which had severally an attraction for the two opposite poles of the electric current. This idea he never lost sight of throughout the whole course of his experiments, though he repeatedly shifted his ground in regard to the contrivances by which he sought its proof and manifestation. To proceed in any other way would not be to philosophise, but merely, as it were, to dip the hand into the bag of chance in quest of a discovery, as men draw prizes at a lottery. It is true that, until the experiment has confirmed or refuted his expectations, this guiding idea upon which the experimenter proceeds must be regarded merely as a conjecture. But such a conjecture or hypothesis he must have in his mind, or he is in no condition to set about the inquisition of nature. What progress would the conductor of a trial in a court of justice be likely to make, in questioning a witness, without some previous notion of the truth which the evidence was likely to establish? He might waste the whole day in putting questions and receiving answers, and at last have ascertained nothing. Just as unprofitably would the interrogator of nature spend his time if he had no directing anticipation in every case, according to which to order his experiments. Accident might, it is possible, throw a discovery in his way; but his own occupation would be evidently as idle and as little that of a philosopher as the rattling of a dice-box. *Whenever, indeed, a discovery is made without being anticipated, we say that it has been made by chance.* On the other hand, the history of all discoveries that have been arrived at by what can with any propriety be called philosophical investigation and induction, attests that necessity which has been asserted of the experimenter proceeding in the institution and management of his experiments upon a previous idea of the truth to be evolved. This previous idea is what is properly called a *hypothesis*, which means something *placed under* as a foundation or platform on which to institute and carry on the process of investigation. A *theory* is a completed view of a harmonious system of truths, evolved and proved by calculation or induction. As the latter is the necessary completion of every philosophical inquiry, so the former is its equally indispensable beginning. It is the aim in the mind of the philosopher, without which he cannot philosophise. It makes, in short, the main difference between the experiments of the philosopher in his laboratory, and those of the child among his play-things. Of course, however, every hypothesis must give way before an experiment the result of which cannot be reconciled with it. Newton, in proceeding to investigate the system of the heavens, set out on the hypothesis that the same power of gravitation which made a stone fall to the ground would be found to retain the moon and the planets in their orbits around the earth and the sun. The result of his first calculation was unfavourable to this supposition, and he at once abandoned it. We have here an example both of the use of an hypothesis, and of the proper limits of reliance on it. The grand discovery which eventually resulted from Newton's investigations affords us, again, an illustration of the manner in which an hypothesis serves to lead to, and originate a theory.

The metal which Sir Humphrey Davy obtained from potash he called *Potassium;* and from soda he also, by a similar process, obtained another, which he called *Sodium.* Both these new metals he found to possess several curious properties, which, however, we cannot here stop to enumerate. He afterwards decomposed also the different earths, and showed them to be all, as well as the alkalis, compounds of oxygen with a metallic base. But these important discoveries, which may be said to have revolutionised the science of chemistry, were not the only results which he obtained from his galvanic and electrical experiments. The interesting subject of the connection between electricity and magnetism received considerable elucidation from his researches. For an account of his contributions to this branch of science, we must refer to the able memoir we have already mentioned, or to his papers on the magnetic phenomena produced by electricity, in the Philosophical Transactions.

Meanwhile his attention had been attracted to another subject of the greatest practical importance—the possibility of preventing the destructive explosions in coal-mines occasioned by the fire-damp, or inflammable gas, which is found in many parts of them. By a series of experiments, Davy found that this dangerous gas, which was known to be nothing more than the hydrogen of the chemists, had its explosive tendencies very much restrained by being mixed with a small quantity of carbonic acid and nitrogen (the ingredient which, along with oxygen, forms atmospheric air); and that, moreover, if it did explode when so mixed, the explosion would not pass through apertures less than one-seventh of an inch in diameter. Proceeding, therefore, upon these ascertained facts, he contrived his celebrated *Safety Lamp*. It consists of a small light, fixed in a cylindrical vessel, which is everywhere air-tight, except in the bottom, which is formed of fine wire gauze; and in the upper part, where there is a chimney for carrying off the foul air. The air admitted through the gauze suffices to keep up the flame; which, in its combustion, produces enough of carbonic acid and nitrogen to prevent the fire-damp, when inflamed within the cylinder, from communicating the explosion to what is without. The heretofore destructive element, thus caught and detained, is therefore not only rendered harmless, but actually itself helps to furnish the miner with light, the whole of the interior of the cylinder being filled with a steady green flame, arising from the combustion of the hydrogen, which has been admitted to contact with the heat, but cannot carry back the inflammation it has received to the general volume without. Armed with this admirable protection, therefore, the miner advances without risk, and with sufficient light to enable him to work, into recesses which formerly he could not have dared to enter. The safety lamp has been the means of saving many lives, and has enabled extensive mines, or portions of mines, to be wrought, which but for its assistance must have remained unproductive. The coal-owners of the northern districts invited Sir Humphrey, in 1817, to a public dinner, and presented him with a service of plate of the value of £2000, in testimony of what they felt to be the merit of this invention.

We will mention only another of this eminent individual's ingenious practical applications of those scientific truths with which he enriched the philosophy of his age. About the year 1823, the attention of the Commissioners of the Navy was so strongly excited to the fact of the rapid decay of the copper sheathings of ships when exposed to the action of the salt-water, that they applied to the Royal Society to take the subject into consideration, and endeavour to devise a remedy for the evil. On this occasion, Davy again had recourse to those principles of electro-chemistry, of which he had himself been the discoverer, and by the application of which he had already obtained so many brilliant results. One of the laws of electrical agency which he considered himself to have ascertained, was that two substances can only combine by what is called chemical affinity or attraction when they are in opposite electrical states,—that is to say, when the one is positively, and the other negatively, electrified. The copper and the water, therefore, he concluded, were naturally in these circumstances; and all that would be required, consequently, to prevent the action of the one upon the other, would be to change the electrical condition of that one of them—namely, the copper—which it was possible to submit to the necessary treatment. He thought of various ways of effecting this object; but, at last, he determined to try the effect of merely placing a quantity of zinc or iron in contact with the copper; the former metals being more positive than the latter, and therefore fitted by induction to repel a portion of its electricity, and so to render it negative like the water. The result surpassed his expectations. So powerfully did the one metal act in reversing the electrical state of the other, that a bit of zinc or iron, no larger than a pea, was found sufficient to protect from corrosion forty or fifty square inches of copper. Nothing, therefore, could be more perfect than the success of this contrivance for the particular purpose it was intended to serve. But, unfortunately, it has been found by experience, that, although Davy's method completely answers for preventing the wasting of the copper, the sea-weeds and marine insects accumulate in such quantities upon the bottoms of ships so protected, that they become, after a short time, scarcely navigable. At the time, therefore, the use of the zinc and iron was of necessity abandoned. It is by no means improbable, however, that some expedient may be contrived for counteracting this consequence of the application of Davy's invention—in which case it will be entitled to rank as one of the most valuable discoveries ever made.

We have thus, guided chiefly by the Memoir of which mention has been made above, pursued the principal triumphs of Sir Humphrey's splendid career, and described what he achieved, although cursorily and briefly, in such a manner, we trust, as to put even the unscientific reader in possession of a tolerably just view of the great discoveries on which his fame rests. In 1827, as we have already mentioned, his health had become so bad, that he found it necessary to resign the presidency of the Royal Society. Immediately after this he proceeded to the Continent. During his absence from England, he still continued to prosecute his chemical researches, the results of which he communicated in several papers to the Royal Society. He also, notwithstanding his increasing weakness and sufferings, employed his leisure in literary composition on other subjects, an evidence of which appeared in his "Salmonia," a treatise on fly-fishing, which he published in 1828. This little book is full of just and pleasing descriptions of some of the phenomena of nature, and is imbued with an amiable and contented spirit. His active mind, indeed, continued, it would seem, to exert itself to the last almost with as unwearied ardour as ever. Beside the volume we have just mentioned, another work, entitled "The Last Days of a Philosopher," which he also wrote during this period, was given to the world after his death. He died at Geneva on the 30th of May, 1829. He had only arrived in that city the day before; and having been attacked by apoplexy after he had gone to bed, expired at an early hour in the morning.

No better evidence can be desired than that we have in the history of Davy, that a long life is not necessary to enable an individual to make extraordinary advances in any intellectual pursuit to which he will devote himself with all his heart and strength. This eminent person was, indeed, early in the arena where he won his distinction; and the fact, as we have already remarked, is a proof how diligently he must have exercised his mental faculties during the few years that elapsed between his boyhood and his first appearance before the public, although, during this time, he had scarcely any one to guide his studies, or even to cheer him onward. Yet notwithstanding that, he had taken his place, as we have seen, among the known chemists of the age almost before he was twenty-one, the whole of his brilliant career in that character, embracing so many experiments, so many literary productions, and so many splendid and valuable discoveries, extended only over a space of not quite thirty years. He had not completed his fifty-first year when he died. Nor was Davy merely a man of science His general acquirements were diversified and extensive. He was familiar with the principal continental languages, and wrote his own with an eloquence not usually found in scientific works. All his writings, indeed, show the scholar, and the lover of elegant literature, as well as the ingenious and accomplished philosopher. It not unfrequently happens that able men, who have been their own instructors, and have chosen for themselves some one field of exertion in which the world acknowledges, and they themselves feel, their eminence, both disregard and despise all other sorts of knowledge and acquirement. This is pedantry in its most vulgar and offensive form; for it is not merely ignorant, but intolerant. It speaks highly in favour of the right constitution and the native power of Davy's understanding, that educated as he was, he escaped every taint of this species of illiberality; and that while, like almost all those who have greatly distinguished themselves in the world of intellect, he selected and persevered in his one favourite path, he nevertheless revered wisdom and genius in all their manifestations.

GEORGE STENSON.

A RAILWAY train, a steamer, or our complicated system of telegraphy, are all often triumphantly pointed to as the high-water mark of modern civilisation. While there is nothing wrong in this, it is at the same time self-evident that it is well to bear the thought in mind, that our boasted progress has been a slow growth, that a thousand lives and influences have been used to help it forward, and that the men of the present age are laying down or recreating the foundations of the life and wellbeing of the generations to come. When we think of railways and the locomotive, the names of James Watt and George Stephenson rise to mind. We intend briefly and simply to trace the principal incidents in the career of the latter. To all true workers it is full of stimulus and encouragement, for work done wisely, worthily, and well, anywhere, unites the worker to that great fellowship, known or unknown to the world, who labour as in the sight and hope of heaven, and do well the little common duties of every day.

George Stephenson was born on the 9th of June, 1781, at the village of Wylam, about eight miles west of Newcastle-on-Tyne. His parents were poor but respectable, and his father, Robert Stephenson, was fireman of the pumping-engine at Wylam, and he is described by Dr. Smiles as of an amiable disposition. While tending his engine fire he would draw around him the young folks of the village and tell them the story of Sinbad the Sailor, or Robinson Crusoe. He was partial to birds, and would sometimes go bird-nesting, and once took young George to see a nestful of young blackbirds, a sight which he never forgot. None of Stephenson's children went to school, as his limited income would not admit of it. The common two-storied, red-tiled building, where they dwelt, stood just beside the wooden tramway on which the coal-waggons were drawn by horses from the coal-pit to the loading-quay, and one of the duties of the elder children was to watch and keep the younger ones out of the way of the waggons, which were daily dragged up and down by horses. Eight years of his life had passed when the Stephenson family removed to Dewley Burn. Young Stephenson's first actual employment was to herd a neighbour's cows at the wage of twopence a-day. Like other boys of his age, he spent much of his time in bird-nesting, in making whistles, and in erecting little water-mills in the streams near by. "But his favourite amusement at this early age was in erecting clay engines in conjunction with his chosen playmate, Tom Thirlaway. They found the clay for their engines in the adjoining bog; and the hemlock which grew about, supplied them with abundance of imaginary steam-pipes." His next work was to lead the horses when plough-ing, or to hoe turnips and other farm work.* When taken on at the colliery and employed to clear the coal of stones, bats, and dross, his wages were advanced to sixpence a-day, and afterwards to eightpence, when he was set to drive the gin-horse. While driving the gin at Black Callerton Colliery, two miles from Dewley Burn, he indulged his fondness for bird-nesting in the hedgerows as he passed along to and from his work. He also indulged himself in a stock of tame rabbits, and used to pride him on the superiority of his breed. When fourteen he was appointed assistant fireman to his father at Dewley Burn, at the wage of one shilling a-day. His great ambition at the time was to become an engineman. When his wages were afterwards raised to twelve shillings a-week at another place, on announcing the fact to his fellow-workmen, he added, "I am now a made man for life." At seventeen, he had charge of a pumping engine at which his father acted as fireman. His duty was to watch the engine and see that it worked well, also that the pumps were drawing efficiently. But Stephenson was no mere mechanical engineman, he applied himself to the study of its different parts, taking it down and putting it together again, so that he was soon able to dispense with the assistance of the engineer of the colliery. When eighteen years of age, and when in charge of the engine at a wage of twelve shillings a-week, he began to remedy his defective education, and commenced to learn to read. His first teacher was Robin Cowens, a poor teacher in Walbottle, who kept a night school, which was attended by a few of the colliers' and labourers' sons in the district. This school was exchanged for one kept by a Scotch dominie, where Stephenson made rapid progress in arithmetic, and also learned to write. When twenty years of age he had become brakesman at Black Callerton pits. The duty of the brakesman was to superintend the working of the engine and machinery by means of which the coals were drawn out of the pit. He also took his turn on the night-shift, and his vacant night hours were either utilised in doing sums, in practising writing, or in shoemaking or mending. An attachment formed at this time for a young woman named Fanny Henderson, a servant at a neighbouring farm-house, stimulated him in the extra efforts he was making. He married Fanny Henderson when he was twenty-one; by thrift, sobriety, and industry he was enabled to take up house, although in very humble style, at Willington Ballast Quay. The marriage took place on the 28th November, 1802. At his daily work Stephenson continued to study the principles of mechanics, and to master the laws by which his engine worked. His evenings spent at home beside his young wife were always turned to some account. An attempt to discover perpetual motion, although it did not succeed, certainly helped to awake his inventive faculties. Having taking his own clock to pieces and cleaned it, and this becoming known in the neighbourhood, he soon had plenty to do in the same line.

His only son Robert was born at Willington Quay on the 16th October, 1803. After working for about three years as a brakesman at Willington, he removed, in 1804, to West Moor Colliery, Killingworth, to a similar situation. Killingworth lies about seven miles north of Newcastle, and it was here that his practical ability as a workman and engineer began to be recognised by his employers. Shortly after his settlement in his new home, to his great sorrow, his wife died. An engagement to superintend the working of one of Boulton & Watt's engines at Montrose, in Scotland, took him away from Killingworth for about a year. On his return he had greatly increased his practical knowledge, and had also saved about £28. The journey both going and returning was accomplished on foot. He found his father had met with an accident, and was in great poverty. George Stephenson paid off his debts and made provision for him. His prospects during the years 1807-8 were very discouraging. Great Britain was engaged in a war; the necessaries of life were heavily taxed; and finally, he was drawn at that time for the militia. He found a substitute, however, by paying a certain sum. So down-hearted was he at this time that he meditated emigrating to the United States. An opportunity was not long in occurring, which, being taken advantage of, materially helped him forward. The engineers of that time, as Mr. Smiles tells us, worked very much in the dark, and, for the most part, without any knowledge of the principles of mechanics. An atmospheric or Newcomen engine, made by Smeaton, had proved a failure, and no engineer or workman could put it right. A speech which Stephenson had made being reported to the head viewer of the pit as to his ability to "alter her and make her draw," and he being dead beat at the time, he at once entrusted him with the work. "Well, George," said the viewer, " they tell me you think you can put the engine at the High Pit to rights." " Yes, sir," said George, " I think I could." "If that's the case, I'll give you a fair trial; and you must set to work immediately. We are clean drowned out, and cannot get a step further. The engineers are all beat, and if you really succeed in accomplishing what they cannot do, you may depend upon it I will make you a man for life." The repairs occupied Stephenson about four days, and were done, if roughly, yet on scientific principles; and before the end of the same week the pit was so far clear of water that the miners could be sent to the bottom. For the successful accomplishment of this work, Stephenson received the sum of £10, with which he was highly gratified. In addition, he was appointed engineman at the High Pit on good wages. His success in doctoring the engine led to his being very extensively consulted by the owners of wheezy and ineffectual pumping-machines in the neighbourhood; and in his treatment of them he is said to have left the regular engineers far behind.

Robert Stephenson was meanwhile receiving as good an education as his father could afford. After the village school of Long Benton had done something for him, he was sent to Bruce's Academy, Newcastle, to which place he rode backwards and forwards on a donkey. To his home, near West Moor Pit, Killingworth, which originally consisted of but one apartment on the ground floor, Stephenson gradually added until he made it a comfortable four-roomed dwelling. In the garden attached, it was his pride to cultivate gigantic leeks and large cabbages. In his leisure time he was still fond of displays of feats of strength and agility, at which he often distanced his competitors. In 1812, the engine-wright at Killingworth was killed by an accident, when George Stephenson was promoted to his post at a salary of £100 a-year. This relieved him from the routine of manual labour, but his brain and his hands were kept as busy as ever. The first winding-engine for drawing the coals out of the pit, and the first pumping-engine erected by him for Long Benton Colliery, were both successful. In some evidence which he gave before a select committee of the House of Commons in 1835, he thus spoke of his life at this time: "After making some improvements in the steam-engines above ground, I was then requested by the manager of the colliery to go underground along with him to see if any improvement could be made in the mines by employing machinery as a substitute for manual labour and horse-power in bringing the coals out of the deeper workings of the mine. On my first going down to Killingworth Pit, there was a steam-engine underground for the purpose of drawing water from a pit that was sunk at some distance from the first shaft. The Killingworth coal-field is considerably dislocated. After the colliery was opened, at a very short distance from the shaft, one of these dislocations was met with. The coal was thrown down about forty yards. Considerable time was spent in sinking another pit to this depth. And on my going down to examine the work, I proposed making the engine (which had been erected some time previously) to draw the coals up an inclined plane, which descended immediately from the place where it was fixed. A considerable change was accordingly made in the mode of working the colliery, not only in applying the machinery, but employing putters instead of horses in bringing the coals from the hewers; and by those changes, the number of horses in the pit was reduced from about 100 to 15 or 16. During the time I was engaged in making these important alterations, I went round the workings in the pit with the viewer almost every time that he went into the mine, not only at Killingworth, but at Mountmoor, Derwentcrook, Southmoor, all of which collieries belonged to Lord Ravensworth and his partners; and the whole of the machinery in these collieries was put under my charge." The fact of his son Robert being a member of the Newcastle Literary and Philosophical Institution was of some assistance to his father. He brought home books, or failing that, drawings from scientific articles, which were talked over between the father and son at home, and made to conduce to the improvement of both.

Before we come to the details of George Stephenson's improvements on the locomotive, it might be well to glance at

THE EARLY HISTORY OF THE RAILWAY.

Two centuries ago, in the life of Lord Keeper Guildford, we read the following: "When men have pieces of ground between the colliery and the river, they sell *leave* to lead coals over their ground, and so dear, that the owner of a rood of ground will expect £20 per annum for this leave. The manner of the carriage is by laying rails of timber from the colliery down to the river, exactly straight and parallel; and bulky carts are made with four rowlets fitting these rails, whereby the carriage is so easy that one horse will draw down four or five caldrons of coals, and is an immense benefit to the coal merchants."

There is mention made of tramways as early as 1602; but there is some convenience in accepting the period of two centuries as the starting-point in noticing the history of railways. The tramways described in the above extract were of wood, and it was not till the opening of the eighteenth century that the wood came to be protected with iron. In the early part of that century many tramways appear to have been laid down to connect collieries with the ports whence the coal was shipped. One of these has obtained some historical interest; namely, the railway between Tranent colliery and its port of Cockenzie, in East Lothian—a railway still in existence—part of the embankment of which was used as a position for his cannon by "Johnny Cope" in the battle of Prestonpans in 1745. In the travels of St. Fond it is mentioned that coals could be imported from England at Marseilles cheaper than French coals of inferior quality, and the facilities for conveying coals to the ports in this country, by the use of the tramways, and the method of shipping direct from the waggons, is believed to have had some share in bringing about this result.

One of the earliest records of the use of iron to protect the wooden trams is in connection with the ironworks at Colebrookdale, in Shropshire, subsequently celebrated for the erection of the first considerable iron bridge, and where, about 1760, iron plates were nailed to the wooden rails, as well to diminish friction as to prevent abrasion. This soon led to the substitution of rails of solid iron, which was attended with rapid success, and adopted in various parts of the country. There was, for instance, a railway five miles long, from the collieries in the vicinity of Derby into that town; there was another called the Park Forest Railway, about six miles long; and another, near Ashby-de-la-Zouch, in Leicestershire, which had four miles of double and eight miles of single rails. Towards the beginning of the present century, railways had made their way into all coal and mining districts, and their progress was so rapid that in 1811 there were in South Wales not less than 150 miles of railways, of which the Merthyr-Tydvil Company possessed thirty miles.

Amongst personal reminiscences of these primitive railways by persons living at the time, it may be interesting to quote those of Mr. Robert Reid, who was born in 1772. In his interesting memoirs of "Old Glasgow," he says: "I remember the coal quay, which stood at the present ferry, west end of Windmill Croft. It was built by the Dumbarton Glass Work Company to convey coals from the lands of Little Govan to their works at Dumbarton. The river was then deeper at the coal quay than at the Broomielaw There was a timber tramway from the Little Govan works to the said quay, which ran through the lands of Kingston, and by the road on the east side of Springfield. *I have walked upon this tramroad, which I believe was the first of our Glasgow railways.* The Dumbarton Glass Work Company also possessed a tramroad on the north side of the Clyde, from the coal works in the neighbourhood of Gartnavel."

But while in regard to the transit and shipment of coals this considerable advance was made, the other branches of traffic, depending on the wretched country roads of last century, remained for half-a-century longer in the depths of barbarity.

"I observed to-day," says Boswell, in his "Tour to the Hebrides," "that the common way of carrying home their grain here is in loads, on horseback. They have also a few sleds or *cars*, as we call them in Ayrshire, clumsily made and rarely used." An aged East Lothian farmer, recently dead, informed the writer that in his youth the mode of bringing grain to the market at Haddington was on pack-horses. This was within recent memory before there were either made roads or railways!

The solid iron rails mentioned as having been introduced at Colebrookdale were called "scantlings," and consisted of five feet long pieces, four inches in breadth, which were laid down under the wheel, simply to decrease friction, as the wooden trams had previously been. The next stage, that of casting rails with an upright flange to keep the wheels on the track, was reached about 1776, in connection with a colliery belonging to the Duke of Norfolk, near Sheffield. Though the flange was subsequently taken from the rail and put on the wheel, the first century of railway history closes with the adoption of the two chief features of the railway as a travelling track—the use of cross sleepers on which to fasten the rails, and the introduction of the flange to keep the cars upon the track.

A quarter of a century brought the invention of the oval rail, with a grooved tire upon the wheels, another step towards the realisation of subsequent success. This "edge railway," as it was called, was first used at Lord Penrhyn's slate quarries in Wales. It being found that the oval rail wore into the wheel and caused it to stick, the next step was to make the surface of the rail and the edge of the wheel flat, and, *voila tout*, the rail way, as we know it, was made. There have been many improvements in the mode of manufacture, in the kinds of sleepers used (stone or wood), in the method of fastening them, in the introduction of steel rails; in the discovery, very recently, that iron rails can be made even more durable and less expensive than steel. But the fundamental condition of the rail remains unchanged, and on the plan thus introduced early in the century all our great progress of to-day has been made.

Mr. R. L. Edgeworth, writing in *Nicholson's Journal of the Arts*, in 1802, describes a project formed by him many years before for laying iron railways for baggage waggons on the great roads of England. Objections as to first cost and maintenance had deterred him from promoting it, and to obviate the latter he proposed to use a series of smaller cars—the modern "train"—in order to save the wear of the rails. In 1768 he obtained the Society of Arts' gold medal, for models of his carriages, and twenty years later he made four carriages which were used for some time on a wooden line of rails to convey lime for farming purposes. Besides using his proposed railways for heavy waggons at a slow pace, Mr. Edgeworth thought means might be found of enabling stage-coaches to go *six* miles an-hour, and post-chaises and gentlemen's travelling carriages at *eight* miles an-hour, both with one horse. Another proposal he made was that small (stationary) engines placed from distance to distance might by means of circulating chains be made to draw the carriages along roads with a great diminution of horse labour and expense.

An attempt to take a systematical commercial view of the utility of railways was made in 1800, by Dr. James Anderson, in the fourth volume of his "Recreations in Agriculture." He proposed to construct railways by the side of the turnpike roads, so as to follow the ordinary levels and lines of traffic : to commence with the highway from London to Bath. Where the road ascended a hill, the level was to be sought by going round its base, constructing a viaduct, or piercing a tunnel; and so carefully were these contingencies discussed, that, with the exception of horses being the moving power, his plans and arguments might be accepted as the description of a railway of the present day. One point particularly insisted on was, that the railways should be managed by Government, not by private companies, who would unite monopoly with speculation; but should "be kept open and patent to all alike who shall choose to employ them, as the king's highway, under such regulations as it shall be found necessary to subject them by law." No immediate result followed the publication of Dr. Anderson's views; no one had then thought of railways independent of other thoroughfares, and to border the latter by iron routes was not to be entertained.

There is another name connected with the rise of railways which cannot be left unnoticed—Thomas Gray, of Leeds. Hearing, while on the Continent in 1816, that a canal had been projected to connect the coal-fields of Belgium with the frontier of Holland, he recommended the making of a railway instead. His mind had been for some time directed to the subject; and in 1818 he showed to his friends a manuscript containing observations on a railroad for the whole of Europe. Soon after he returned to England for the purpose of making his scheme public, and in 1820 he published a volume entitled "Observations on a General Iron Railway, or Land Steam Conveyance, to supersede the Necessity of Horses in all Public Vehicles : Showing its vast Superiority in every respect over the Present Pitiful Methods of Conveyance by Turnpike Roads and Canals." In this work, among advantages to result from the new system, Gray showed that fish, vegetables, agricultural and other perishable produce might be rapidly carried from place to place; that two post deliveries in the day would be feasible; and that insurance companies would be able to promote their own interests by keeping railway fire-engines, ready to be transported to the scene of a conflagration at a moment's warning. The cost of construction was calculated at £12,000 a-mile; and his plan included a trunk line from London to Plymouth and Falmouth; lines to Portsmouth, Bristol, Dover, and Harwich; an offset from the latter to Norwich, a trunk line from London to Birmingham and Holyhead, another to Edinburgh by Nottingham and Leeds, with secondary lines from Liverpool to Scarborough, and from Birmingham to Norwich. His system was not only remarkable for its simplicity, but comprehended all the important towns of the kingdom, and was in many respects preferable to the lines subsequently made. His plan for Ireland had a grand trunk line from Dublin to Derry, another to Kinsale, and by lesser lines ramifying from these he sought to connect all the chief towns with the Irish capital. Regarding his projects, Sir John Hawkshaw, in his British Association speech (1875), remarked: "No sooner had our ancestors settled down with what comfort was possible in their coaches, well satisfied that twelve miles an-hour was the maximum speed to be obtained or that was desirable, than they were told that steam conveyance on iron railways would supersede their 'present pitiful' methods of conveyance. Such was the opinion of Thomas Gray, the first promoter of railways, who published his work on a general iron railway in 1819. Gray was looked on as little better than a madman. 'When Gray first proposed his great scheme to the public,' said Chevalier

Wilson, in a letter to Sir Robert Peel in 1845, 'people were disposed to treat it as an effusion of insanity.' The struggles which preceded the opening of the first railway were brought to a successful issue by the determination of a few able and far-seeing men; and the names of Thomas Gray and Joseph Sandars, of William James and Edward Pease, should always be remembered in connection with the early history of railways, for it was they who first made the nation familiar with the idea."

Whatever effect Gray's persevering labours may have had in directing attention to the subject of railways, he himself gained neither reward nor honour. His late years were passed in obscurity as a dealer in glass on commission at Exeter, in which city he died in October, 1848, at the age of sixty-one. He died, it is said, "steeped to the lips in poverty."

In an early number of *Blackwood's Magazine* we have a notice of a railway in Munich nearly contemporary with the proposals of Gray: "We have received a report from Munich, which, if it be not exaggerated, well deserves the attention of our countrymen. A model, on a large scale, of an iron railroad, invented and completed by the chief counsellor of the mines, Joseph von Baader, has been received at the Royal Repository for Mechanical Inventions, which is said to surpass in utility whatever has been seen in England; some say by a proportion of two-thirds, although it costs less by half. On a space perfectly level, laid with this invention, a woman or a child may draw with ease a cart laden with fifteen or sixteen hundred-weight. And if no greater inclination than six inches and a-half on a hundred feet in length be allowed, the carts will move of themselves, without any external impulse. A single horse may be the means of conveying a greater weight than twenty-two horses of the same strength on the best of common roads."

While there was thus a gathering together of testimony as regards the improvement of the roads over which wheeled vehicles were to be drawn, there was gradually being developed the idea of employing another and more powerful agent for the propulsion of the vehicles. There cannot be the least doubt that the numerous attempts to apply steam to navigation acted on the minds of men of skill and invention in order to have the same powerful agent applied to the ordinary requirements of the road. Indeed, the first invention of William Symington was applied to a carriage as well as to a barge, and his diagram and detail of a steam-carriage were contemporary with his invention of a steamship. It was probably in the knowledge that such ideas were being wrought out into practical shape that the lines were written by Dr. Darwin, to which the reputation of prophecy has almost attached:

> "Soon shall thy arm, unconquered steam ! afar
> Drag the slow barge, or drive the rapid car ;
> Or on wide-waving wings expanded bear
> The flying chariot through the fields of air !"

So wrote Dr. Darwin in his *Botanic Garden* in 1793, and the vision of the "flying chariot" does not appear to-day much more extravagant than did, when these lines were published, the prediction of "rapid" travelling by means of a steam-engine. Yet, nearly a century before, a very fair attempt at the construction of a locomotive steam-engine had been made. The scene of the experiment was Japan, and the actors in it were the Jesuit missionaries, who sought to find favour with the Emperor Kanghi. They caused a waggon of light wood to be made, in the middle of which they placed a brazen vessel full of live coals, and on them an "colipile," the wind from which issued through a little pipe upon a sort of wheel made like the sail of a windmill. This little wheel turned another with an axle-tree, and by that means the waggon was set a-running for two hours together. This description is rather that of a hot-air engine than a steam-engine, but it was a locomotive, and is the earliest of its race.

In the *Conservatoire des Arts et Métiers* in Paris is preserved the steam-carriage constructed by M. Cugnot in 1763, which was a remarkable machine, like a long brewer's cart, with a boiler and engine at one end. It went with such force that it knocked down a wall, and its power was in consequence considered too great for ordinary use, and it was put aside as a dangerous invention.

A model of a steam-carriage was made in 1784 by William Murdoch, the friend and assistant of Watt, but it was of very diminutive proportions.

The suggestion for such an application of steam had been made by Dr. Robison, of Edinburgh, in 1759, to James Watt, who included the idea in his fourth patent, but seems to have doubted the safety of the carriage. He mentioned the idea to Murdoch, who proved practically, on a small scale, the correctness of the calculations that had been made. Of Murdoch's machine, it is narrated that on a dark night in the year named, the venerable clergyman of Redruth, in Cornwall, when walking in a lonely lane leading to his church, heard a most unearthly noise, and beheld approaching him, at great speed, an indescribable creature, glowing with internal fires, and whose gasps for breath seemed to denote some internal struggle of a deadly kind. His cries brought the inventor, William Murdoch, to his side, who explained to him that this terrible monster was nothing more or less than a locomotive he had invented, and which had broken away from his control.

An equal amount of terror was created in some minds by the steam-carriage of Richard Trevithick, an eccentric engineer connected with the Cornish tin-mines, who had seen Murdoch's small carriage. In 1802 he took out a patent for this novel machine, which was exhibited to large crowds of spectators on what is now the site of Euston station. Coleridge relates that, when it was being conveyed from the place in Cornwall where it was constructed to the port at which it was shipped to London, after carrying away a portion of the rails of a gentleman's garden, it came in sight of a closed toll-gate. Trevithick immediately shut off the steam, but the momentum was so great that the carriage proceeded some distance, coming dead up, however, just on the right side of the gate, which was opened like lightning by the gatekeeper. "What have us got to pay here?" asked Trevithick's cousin, Andrew Vivian, who accompanied him. The poor toll-man, trembling in every limb, his teeth chattering in his head, essayed a reply: "Na, na, na, na." "What have us got to pay, I say?" "No—noth—nothing to pay! My dear Mr. Devil, do drive on as fast as you can! Nothing to pay!"

Trevithick constructed another steam-carriage for railway purposes, which, in 1804, ran on the Merthyr-Tydvil tramway in South Wales. It drew a load of ten tons at the rate of five miles an-hour.

The earliest locomotives were designed to run upon a perfectly smooth line and a straight road, and for many years it was supposed that they could not climb hills or be made to go round corners unless the wheels were provided with a cogged rim to work on a corresponding rack along the rails. The cogged or toothed wheels and rails were introduced in 1811 by Mr. Blenkinsop, of Leeds. It was not till 1813 that Mr.

Blackett, of Wylam, a coal proprietor, established the fact that locomotives, running with smooth wheels on smooth rails, could draw heavy loads up a moderate incline. His engine, called the "Puffing Billy," was otherwise clumsily constructed. It had only a single cylinder, and was full of pumps, plugs, and other gear, which were always getting out of order.

Lord Ravensworth, the principal partner in the Killingworth Collieries, when the subject of constructing a travelling engine was brought before him and his partners by Stephenson in 1813, empowered him to proceed. "The first locomotive that I made," said Stephenson in a speech made many years afterwards, "was at Killingworth Colliery, and with Lord Ravensworth's money. Yes, Lord Ravensworth and partners were the first to entrust me with money to make a locomotive engine. That engine was made thirty-two years ago, and we called it 'My Lord.' I said to my friends, there was no limit to the speed of such an engine if the works could be made to stand it." It was tried on the Killingworth railway on the 25th of July, 1814. "Blucher," as this locomotive was popularly called, was not eminently successful until the introduction of the steam-blast. At first it drew a load of thirty tons at the rate of four miles an-hour. His second engine, patented in 1815, doubled this speed. "Thus," as Dr. Smiles remarks, "Mr. Stephenson, by dint of patient and persevering labour— by careful observation of the works of others, and never neglecting to avail himself of their suggestions—had succeeded in manufacturing an engine which included the following important improvements on all previous attempts in the same direction, viz.: simple and direct communication between the cylinder and the wheels rolling upon the rails; joint adhesion of all the wheels, attained by the use of horizontal connecting rods; and finally, a beautiful method of exciting the combustion of the fuel by employing the waste steam, which had formerly been allowed uselessly to escape into the air. Although many improvements in detail were afterwards introduced in the locomotive by Mr. Stephenson himself, as well as by his equally-distinguished son, it is perhaps not too much to say that this engine, as a mechanical contrivance, contained the germ of all that has since been effected. It may in fact be regarded as the type of the present locomotive engine."

Explosions of fire-damp frequently took place in Stephenson's time, which were as disastrous in their results as those of more modern times. In the year 1814 an alarm was raised at one of the Killingworth mines, that one of the deepest mains was on fire. Stephenson coming to the pit-mouth ordered the engine-man to lower him down the shaft. Getting six men to volunteer to follow him, he speedily ran up a wall at the entrance to the main, which extinguished the fire and saved many in the mine from a violent death. One of the men, by name Kit Heppel, asked him at this time, "Can nothing be done to prevent such awful occurrences?" Stephenson replied that he thought something might be done. "Then," said Heppel, "the sooner you start the better; for the price of coal-mining now is *pitmen's* lives." This set him a-thinking and working, and in the course of 1815 he endeavoured to give his idea of a miner's safety-lamp a practical shape. He described this lamp to the Committee of the House of Commons, when sitting on the subject of Accidents in Mines. He began by saying that he knew nothing of chemistry at the time. "Seeing the gas lighted up, and observing the velocity with which the flame passed along the roof, my attention was drawn to the contriving of a lamp, seeing it required a given time to pass over a given distance. My idea of making a lamp was entirely on mechanical principles; and I think I shall be found quite correct in my views from mechanical reasoning. I knew well that the heated air from the fire drove round a smoke-jack, and that caused me to know that I could have a power from it. I also knew very well that a steam-engine chimney was built for the purpose of causing a strong current of air through the fire. Having these facts before me, and knowing the properties of heated air, I amused myself with lighting one of the blowers in the neighbourhood of where I had to erect machinery. I had it on fire; the volume of flame was coming out the size of my two hands, but was not so large but that I could approach close to it. Holding my candle to the windward of the flame, I observed that it changed its colour. I then got two candles, and again placed them to the windward of the flame; it changed colour still more, and became duller. I got a number of candles, and placing them all to the windward, the blower ceased to burn. This then gave me the idea, that if I could construct my lamp so as, with a chimney at the top, to cause a current, it would never fire at the top of the chimney; and by seeing the velocity with which the ignited fire-damp passed along the roof, I considered that, if I could produce a current through tubes in a lamp equal to the current that I saw passing along the roof, I should make a lamp that could be taken into an explosive mixture without exploding externally." After many experiments his third and most successful safety-lamp was constructed, and tested on the 30th November, 1815, before he had ever heard of Sir Humphry Davy's experiments. After being exhibited before the Literary and Philosophical Society of Newcastle, it came into use in the

Killingworth Collieries. Since its introduction no accident is known to have taken place from its use. To distinguish it from the "Davy" lamp it is known as the "Geordy" lamp. Stephenson's claim for the independent invention of a safety-lamp was acknowledged at a public dinner given in the Assembly Rooms at Newcastle, in January, 1818, when he was presented with £1000 and a silver tankard. Dr. Smiles is of opinion that the "Geordy" lamp, when severely tested in the mines by an escape of gas, is decidedly the safer of the two. Cases have occurred where the "Davy" lamp has grown red-hot in an explosive atmosphere, while the "Geordy" was entirely extinguished.

The locomotives constructed by George Stephenson in 1816 were, until lately, working regularly on the Killingworth railway, dragging coal trains at the rate of between five and six miles an-hour. This says much for the thoroughness with which they were manufactured. "There were many highly-educated engineers," writes Dr. Smiles, "living in his day, who knew vastly more than he did—trained as they had been in all the science and learning of the schools; but there was none so apt in applying what they knew to practical purposes as the Killingworth 'brakesman' and 'engine-wright.' The great secret of his success, however, was his cheerful perseverance. He was never cast down by obstacles, but seemed to take a pleasure in grappling with them, and he always rose from each encounter a stronger as well as a wiser man. He knew nothing of those sickly phantasies which men, who suppose themselves to be 'geniuses,' are so apt to indulge in; nor did his poverty or necessities ever impair the elasticity of his character. When he failed in one attempt, he tried again and again, until eventually he succeeded." Speaking at a soirée of the Leeds Mechanics'

Institute, he said: "He had commenced his career on a lower level than any man present there. He made that remark for the purpose of encouraging young mechanics to do as he had done—to persevere. And he would tell them that the humblest amongst them occupied a much more advantageous position than he had done on commencing his life of labour. They had teachers who, going before them, had left their great discoveries as a legacy and a guide; and their works were now accessible to all, in such institutions as that which he addressed. But he remembered the time when there were none thus to guide and instruct the young mechanic. With a free access to scientific books, he knew, from his own experience, that they could be saved much unnecessary toil and expenditure of mental capital."

We now give a *résumé* of railway progress in the United Kingdom, with which Stephenson was afterwards so vitally connected, and which will bring the story of his life up to 1830.

PROGRESS OF THE RAILWAY IN THE UNITED KINGDOM, 1801-1830.

The first Act of Parliament for the construction of a railway was passed in 1801, and was promoted by the Surrey Iron Company, for a railway nine miles long, from Wandsworth to Croydon, with a branch to Carshalton. The capital was £60,000, being about what is now considered the normal cost for a "light" railway or local single line locally promoted — namely, £5500 per mile. This, the first line opened under parliamentary sanction, was completed in 1805; and, in connection with its opening, some very interesting experiments in traction were made. Taking the estimate of the draught of a horse, upon a good road, at fifteen hundred pounds, the party of gentlemen who assembled to witness the testing of the line were enabled to judge practically the advantages offered. Twelve waggons were filled with stones till each waggon weighed three tons, and a horse attached to them drew the load, with apparent ease, a distance of six miles in an hour and three-quarters. In the course of the journey the horse was repeatedly stopped to show that he had the power of starting the load with apparent ease. At each stoppage other waggons were attached, and the men employed on the line, to the number of about fifty, were also directed to mount the waggons. At the end of the journey the entire load was found to have reached rather more than fifty-five tons!

In all, about twenty Acts were passed prior to that of the Stockton and Darlington line, mostly for short lengths, the longest being a line of thirty miles, from Sutton Pool, near Plymouth, to the neighbourhood of Dartmouth prison, the capital of which is stated at the extremely low total of £35,000.

The first line authorised in Scotland was from Kilmarnock to Troon, in Ayrshire, for which an Act was obtained in 1808, and which, for a length of about ten miles, was estimated to cost £65,000. This line was opened in 1810, and forming as it does an integral part of the large system now embraced under the name of the Glasgow and South-Western Railway, it entitles that company to be considered the *premier* railway in Scotland.

It was about this time that Thomas Telford projected a very extensive scheme to connect the east and west of Scotland by a grand line, starting from Berwick, and proceeding by the valley of the Tweed to Kelso, Peebles, and Lanark, to the town of Ayr! "We admire," says the compiler of the "Scottish Railway Shareholders' Manual," in 1849, "the genius and sagacity evinced by so magnificent a design; but we do not wonder that, in face of prejudice and ignorance on the part of the public, in the infancy of railway undertakings, it should have been laid aside and forgotten." In 1811 the Berwick and Kelso Company projected a line occupying part of Telford's ground, but this particular railway was never carried out. The scheme lay dormant for many years—for, unlike more modern Acts, no limitation of time was put in the Act, so that the powers did not lapse—and no step was ever made to carry it out. It is true that later works have occupied nearly all the ground projected by Telford, so that a railway journey by Kelso, Peebles, and Lanark to Ayr, can at this day be made per rail, over substantially the same ground as was taken up by the great engineer. But the main lines of connection for traffic purposes, between the east and west of Scotland, have been found elsewhere.

The greater number of the companies incorporated by the Acts up to this time embraced but few persons, and consisted mostly of merchants or owners of collieries seeking an outlet for their goods. Thus, the capital of the Penrhynmaur line, for which an Act was obtained in 1812, was held by two men, the Earl of Uxbridge and Mr. Holland Griffith. On none of the lines for which Acts were obtained up to 1820 was any other motive-power used or designed than that of horses, and not one of the companies even proposed the adoption of the steam-engine, though the invention was by this time beginning to attract attention, nor did the idea of conveying passengers seem to be entertained.

This was, however, to be completely changed by the Act obtained in 1821, in which the clause defining the method of haulage spoke of "the making and maintaining of the tramroads and the passage upon them of waggons *and other carriages*, with men, and horses, *or otherwise*"—words sufficiently elastic to admit of any power being used. However, taking the Act altogether, it can hardly be considered that Mr. Edward Pease, who was the chief promoter of the line, had it in view to use anything else than horse-power, or that he was much moved by Sir Richard Phillips's recommendation that they should use 'Blenkinsop's steam-engine."

The Stockton and Darlington scheme had to run the gauntlet of a fierce opposition in three successive sessions of Parliament. The application of 1818 was defeated by the Duke of Cleveland, who afterwards profited so largely by the construction of the railway. The ground of his opposition was that the line would interfere with one of his fox covers, and through his influence the Bill was thrown out.

Several energetic men, however, were now at the head of the scheme, and they determined to persevere with it. Amongst these, Edward Pease might be regarded as the backbone of the concern. Opposition did not daunt him, nor failure defeat him. When apparently overthrown, he rose again, like Antæus, stronger than before, and made another and stronger effort. He had in him the energy and patient perseverance of many men.

The next year, 1819, an amended survey of the line was made; and, the Duke of Cleveland's fox cover being avoided, his opposition was thus averted. But as Parliament was dissolved on the death of George III., the Bill was necessarily suspended until another session.

The principal opposition now came from the road trustees, who spread it abroad that the mortgages of the tolls arising from the turnpike road leading from Darlington to West Auckland would be seriously injured by the formation of the proposed railway. On this, Mr. Edward Pease issued a printed notice requesting any alarmed mortgagees to apply to the company's solicitors at Darlington, who were authorised to purchase their securities at the price originally given for them. Tnis notice had the effect of allaying the alarm, and the Bill, though still strongly opposed, was allowed to pass both Houses of Parliament in 1821.

The preamble of the Act sets forth the public utility of the proposed line for the conveyance of coal and other commodities from the interior of the county of Durham to Stockton and the northern parts of Yorkshire. Nothing was said about passengers, for passenger traffic was even then not contemplated; and nothing was said about locomotives, as it was at first intended to work the line entirely by horse-power. The road was to be free to all who chose to place their waggons and horses upon it for the haulage of coal and other merchandise, provided they paid the tolls fixed by the Act.

The company were empowered to charge fourpence a-ton per mile for all coal intended for land sale; but only a halfpenny a-ton per mile for coal intended for shipment at Stockton. The latter low rate was introduced in the Act through the influence of Mr. Lambton, afterwards Earl of Durham, for the express purpose of preventing the line being used in competition against him; for it was not believed possible that coal could be carried at that rate except at a heavy loss. As it was, the low rate thus fixed proved the vital element in the future success of the Stockton and Darlington Railway.

The capital specified by the Act was of small amount, and, as events proved, it was altogether inadequate. The share capital was fixed at £82,000, in shares of £100 each, and, in the event of this not being found sufficient, power was given to raise £20,000 more by shares. If the shares were not taken by the public, then the necessary capital, within the above limits, might be raised by the issue of mortgages or promissory notes. These powers were necessarily greatly enlarged by subsequent Acts to enable the line to be completed and placed in sound working order.

While the Stockton and Darlington Railway scheme was still before Parliament, Mr. Edward Pease was writing articles for a York newspaper, urging the propriety of extending it southward into Yorkshire by a branch from Croft. It is curious now to look back upon the arguments by which Mr. Pease sought to influence public opinion in favour of railways, and to observe the very modest anticipations which even its most zealous advocate entertained as to their supposed utility and capabilities.

"'The late improvements in the construction of railways," Mr. Pease wrote, "have rendered them much more perfect than when constructed after the old plan. To such a degree of utility have they now been brought, that they may be regarded as *very little inferior to canals.*"

"Though the railways at Carron [in Scotland] are not exempt from slight risings and depressions, the reduction which they have occasioned in a distance of six miles merits much attention. Before their establishment the Carron Company paid £1200 monthly on an average for carriage, but since then the number of horses employed has been diminished by three-fourths, and the expenditure on carriage reduced to about £300 a-month, effecting a saving to the company of equal to £10,000 a-year. Coal, lime, stone, and grain can also be conveniently weighed by machines placed under the railway depôts and at different points of loading and discharging. The weighing on departing and arriving would also be a great check to fraud.

"One horse can draw, by means of a railway, on a level or slightly-inclined plane, from eight to sixteen waggons of one ton each, and each waggon may be loaded with different kinds of goods to suit the traffic on the line.

"If we compare the railway with the best lines of common road, it may be fairly stated that in the case of a level railway the work will be increased in at least an eightfold degree. The best horse is sufficiently loaded with three-quarters of a ton on a common road, from the undulating line of its draught, while on a railway it is calculated that a horse will easily draw a load of ten tons. At Lord Elgin's works, Mr. Stephenson, the celebrated engineer, states that he has actually seen a horse draw twenty-three tons thirteen cwt. upon a railway, which was in some parts level, and at others presented a gentle declivity!

"The formation of a railway, if it creates no improvement in a country, certainly bars none, as all the former modes of communication remain unimpaired; and the public obtain, at the risk of the subscribers, another and better mode of carriage, which it will always be to the interest of the proprietors to make cheap and serviceable to the community.

"On undertakings of this kind, when compared with canals, the advantages of which (where an extensive traffic on the ascending or descending line can be obtained) are nearly equal, it may be remarked that public opinion is not easily changed on any subject. It requires the experience of many years, sometimes ages, to accomplish this, even in cases which by some may be deemed obvious. Such is the effect of habit, and such the aversion of mankind to anything like innovation or change. Although this is often regretted, yet, if the principle be investigated in all its ramifications, it will perhaps be found to be one of the most fortunate dispositions of the human mind.

"The discovery of the cast-iron railway is comparatively of recent date. It is not only intimately connected with inland navigation, and originated with it, but will be found, as it becomes more perfect, to add to the efficiency and utility of that system of communication, whilst every step in advance must materially promote the interests of the agriculturist, the miner, the merchant, the mariner, and, in short, of the community at large.

"The system of cast-iron railways is as yet to be considered but in its infancy. It will be found to be an immense improvement on the common road and also on the wooden railway. It neither presents the friction of the tramway nor partakes of the perishable nature of the wooden railway, and, as regards utility, it may be considered as the medium between the navigable canal and the common road. We may, therefore, hope that as this system develops itself our roads will be laid out as much as possible on *one level*, and in connection with the great lines of communication throughout the country."

Such were the modest anticipations of Edward Pease respecting railways, about the year 1818. Ten years after, and an age of progress, by comparison, had been made. Mr. Pease did not at first so much as dream of the locomotive, his anticipations being solely based on the employment of horse-power.

If no other, the Act of 19th April, 1821, had one important and immediate consequence in bringing "the engine-wright of Killingworth" (as George Stephenson modestly styled himself) into contact with Edward Pease. He called at his house, as he told the worthy Quaker, because he had heard of the Act, bringing with him a letter of introduction from the director of the Killingworth pits. The conversation that followed, after George Stephenson had presented his letter to Edward Pease, was highly characteristic of both men. As recorded by Dr. Smiles, the originator of the Stockton-Darlington line "very soon saw that his visitor was the man for his purpose. The whole plans of the railway being still in an undetermined state, Mr. Pease was glad to have the opportunity of gathering from Mr. Stephenson the results of his experience. The latter strongly recommended a railway in preference to a tram-road, in which Mr. Pease was disposed to concur with him. The conversation next turned to the tractive power which the company intended to employ, and Mr. Pease said that they had based their whole calculations on the employment of horse-power. 'I was so satisfied,' said he afterwards, 'that a horse upon an iron road would draw ten tons, for one ton on a common road, that I felt sure that before long the railway would become the king's highway.'

"Mr. Pease was scarcely prepared for the bold assertion made by his visitor, that the locomotive engine with which he had been working the Killingworth railway for many years past was worth fifty horses, and that engines made after a similar plan would yet entirely supersede all horse-power upon railroads. Mr. Stephenson was daily becoming more positive as to the superiority of his locomotive; and on this, as on all subsequent occasions, he strongly urged Mr. Pease to adopt it. 'Come over to Killingworth,' said he, 'and see what my "Blucher" can do. Seeing is believing, sir.' And Mr. Pease promised that on some early day he would go over to Killingworth with his friend John Richardson, and take a look at this wonderful machine that was to supersede horses. On Mr. Pease referring to the difficulties and the opposition which the projectors of the railway had had to encounter, and the obstacles which still lay in their way, Stephenson said to him, 'I think, sir, I have some knowledge of craniology, and from what I see of your head I feel sure that if you will fairly buckle to this railway you are the man successfully to carry it through.' 'I think so too, rejoined Mr. Pease, 'and I may observe to thee, that if thou succeed in making this a good railway, thou may consider thy fortune as good as made.'" The remark and reply were alike characteristic of the promoters of the first railway.

The graphic description, by an early friend, of Mr. Edward Pease, that "he was a man who could see a hundred years ahead," was strikingly proved in the weeks that followed his first interview with George Stephenson. Having accepted the invitation to " come over to Killingworth," and having seen with his own eyes what " my 'Blucher' can do," his mind became at once clear as to the immense future awaiting the introduction of the "iron horse" upon the iron railway, and he not only strongly advocated the use of locomotives, but made himself Stephenson's partner in their manufacture. Through his influence the still unknown engine-wright at Killingworth was appointed engineer of the Stockton-Darlington line, and at his urgent request Mr. Pease applied for a new Act of Parliament giving the Stockton-Darlington Company power to work the railway by means of locomotive engines, and to employ them for the haulage of passengers as well as goods. The Act was obtained with some difficulty, against the bitter opposition of a number of powerful peers, such as the Duke of Cleveland, in the session of 1823, when the construction of the railway, under George Stephenson's supervision, was already going on actively. The first rail of the Stockton and Darlington line had been laid, with considerable ceremony, near the town of Stockton, on the 23rd of May, 1822, and notwithstanding the uninterrupted opposition, frequently growing into acts of personal violence, of hundreds of enemies, backed by the whole mob of the district, the works were pushed on so vigorously, that it was possible to open the line on the day fixed, the 27th of December, 1825—eventful day in railway history, well worthy the great "Jubilee" held at Darlington in 1875.

Of the first interview between Stevenson and Pease, very graphic accounts have been given by Dr. Smiles who had an interview with Mr. Pease in 1854, four years before his death, and when he had reached the patriarchial age of eighty-eight. Hale and hearty, and full of reminiscences of the past, sound in health, with his eye not dimmed or his natural force abated, Mr. Pease narrated many circumstances which the biographer of the engineer has made full use of. He described the appearance of Stephenson as having "an honest, sensible look about him, and so modest and unpretending withal." Stephenson spoke in the strong Northumbrian dialect of his district, and described himself as "only the engine-wright of Killingworth —that's what I am." In the course of the interview, Edward Pease said to Dr. Smiles, with much truth—referring to the growth of the trees in front of his house which he had planted as a boy—"Ay, but railways are a far more extraordinary growth even than these. They have grown up not only since I was a boy, but since I was a man. When I started the Stockton and Darlington, some five-and-thirty years since, I was already fifty years old. Nobody would then have dreamt what railways would have grown to in one man's lifetime."

The 27th day of September, 1825, deserves to be marked as a red-letter day in the calendar of the world's history. On that morning the greatest revolution of modern times was to be inaugurated—the painfully slow development of men's ideas up to that point being followed, though not quite immediately, by results which were none the less consequent upon that day's proceedings, that the persons chiefly engaged in the work failed entirely to see what the future had in store for the world in supplement to the success of the opening.

Tuesday, the 27th of September, 1825, was a great day for Darlington. The railway, after having being under construction for more than three years, was at length about to be opened. The project had been the talk of the neighbourhood for so long that there were few people within a range of twenty miles who did not feel more or less interested about it. Was it to be a failure or a success? Opinions were pretty equally divided as to the railway, but as regarded the locomotive, the general belief was that it would "never answer." However, there the locomotive was—"No. 1"—delivered on to the line, and ready to draw the first train of waggons on the opening day.

A great concourse of people assembled on the occasion. Some came from Newcastle and Durham, many from the Aucklands, while Darlington held a general holiday and turned out all its population. To give *éclat* to the opening, the directors of the company issued a programme of the proceedings, intimating the times at which the procession of waggons would pass certain points along the line. The proprietors assembled as early as six in the morning at the Brusselton fixed engine, where the working of the inclined planes was successfully rehearsed. In this trial, as in the subsequent ceremony, a train of waggons laden with coals and merchandise was drawn up the western incline by the fixed engine in seven and a half minutes, and then lowered down the incline on the eastern side of the hill in five minutes.

In spite of the evil prognostications heard on all sides, the inauguration of the Stockton-Darlington Railway passed over most satisfactorily. The programme issued by the company dated " Railway Office, September 19th, 1825," was as follows :—

"The Stockton and Darlington Railway Company do hereby give notice that the formal opening of this railway will take place on the 27th inst., as announced in the public papers. The proprietors will assemble at the Permanent Steam-engine, situated near the Brusselton Tower, about nine miles west of Darlington, at eight o'clock, and, after examining their extensive inclined plains there, will start from the foot of the Brusselton descending plane at nine o'clock in the following order :—

" 1. The company's locomotive engine.

" 2. The engine's tender, with water and coals.

" 3. Six waggons laden with coals, merchandise, etc.

" 4. The committee, and other proprietors, in the coach belonging to the company.

" 5. Six waggons with seats reserved for strangers.

" 6. Fourteen waggons, for the conveyance of workmen and others.

"The whole of the above to proceed to Stockton.

" 7. Six waggons laden with coals, to leave the procession at the Darlington branch.

" 8. Six waggons, drawn by horses, for workmen and others.

" 9. Ditto.

" 10. Ditto.

" 11. Ditto.

"The company's workmen to leave the procession at Darlington and dine at that place at one o'clock, excepting those to whom tickets are specially given for Yarm, and for whom conveyances will be provided on their arrival at Stockton.

"Tickets will be given to the workmen who are to dine at Darlington, specifying the houses of entertainment.

"The proprietors, and such of the nobility and gentry as may honour them with their company, will dine precisely at three o'clock at the Town Hall, Stockton. Such of the party as may incline to return to Darlington that evening will find conveyances in waiting for their accommodation, to start from the company's wharf there precisely at seven o'clock.

"The company take this opportunity of enjoining on all their workpeople that attention to *sobriety* and *decorum* which they have hitherto had the pleasure of observing.

"The committee give this public notice, that all persons who shall ride upon, or by the sides of the railway, on horseback, will incur the penalties imposed by the Acts of Parliament passed relative to this railway."

Appended to the programme was the following footnote :—

"Any individual desirous of seeing the train of waggons descending the inclined plane from Etherley, and in progress to Brusselton, may have an opportunity of so doing, by being on the railway at St. Helen's, Auckland, not later than half-past seven o'clock."

At the foot of the Brusselton incline the procession of vehicles was formed, consisting of the locomotive engine, "No. 1," driven by George Stephenson himself; after it six waggons loaded with coals and flour, then a covered coach containing directors and proprietors, next twenty-one coal waggons fitted up for passengers (with which they were crammed), and lastly six more waggons loaded with coals.

Strange to say, a man on a horse, carrying a flag, with the motto of the company inscribed on it, *Periculum privatum utilitas publica*, headed the procession ! A lithographic view of the great event, published shortly after, duly exhibits the horseman and his flag. It was not thought so dangerous a place after all. The locomotive was only supposed to be able to go at the rate of from four to six miles an-hour; and an ordinary horse could easily keep ahead of that.

Off started the procession, with the horseman at its head. A great concourse of people stood along the line. Many of them tried to accompany it by running, and some gentlemen on horseback galloped across the fields to keep up with the engine. The railway descending with a gentle incline towards Darlington, the rate of speed was consequently variable. At a favourable part of the road, Stephenson determined to try the speed of the engine, and he called upon the horseman with the flag to get out of the way! Most probably deeming it unnecessary to carry his *Periculum privatum* further, the horseman turned aside, and Stephenson "put on the steam." The speed was at once raised to twelve miles an-hour, and at a favourable part of the road, to fifteen. The runners on foot, the gentlemen on horseback, and the horseman with the flag, were consequently soon left far behind.

Although only George Stephenson's name is mentioned as the driver of the locomotive "No. 1," his principal helper, Mr. William Huntley, who still (1877) lives, deserves some notice. Mr. Huntley, who was born at Acklington in 1798, was ten years in the employment of Stephenson & Co., and superintended the erection of the "first locomotive." On the opening day, he drove it in turn with George Stephenson. Huntley afterwards erected the first locomotive that drew a passenger train in Scotland, in 1831, and since that period he has been connected with the railway system at Dundee. Mr. Huntley has attracted attention by his "continuous grip-brake," for stopping trains at high speed, for which he has decided not to take out a patent, presenting it freely to the consideration of the railway world.

In 1824 an Act was obtained for the construction of the Monkland and Kirkintilloch Railway, from Palace Craig, in Lanarkshire, to the banks of the canal at Kirkintilloch, in Dumbartonshire. The line was completed and opened a year after the Stockton and Darlington, namely, in September, 1826, and was at first used only for the conveyance of coal for shipment at the canal. The waggons were drawn by steam locomotives. As it was found that passenger traffic could be cultivated, the directors, early in 1827, added to most of the coal-trains a coach for the conveyance of passengers, with considerable profit to the company. The Ballochney Railway, opened in 1828, also had some of its trains drawn by locomotives; and here, too, the addition of a passenger carriage proved a source of convenience to the public and of profit to the company.

The Canterbury and Whitstaple was the fourth completed line in the kingdom which used locomotives and carried passengers, and perhaps from the fact that it did not, any more than the two Scotch lines above named, attract much public attention, its story is deserving of record now. The Act for the formation of the line was got in 1825, and with a capital of £35,000, it was proposed to make a line six and a quarter miles long, with heavy gradients, and a tunnel half-a-mile long. This proved inadequate, and subsequent Acts were obtained in 1827 and 1828 to raise new capital. As even these additions proved too little, loans and mortgages were resorted to, and it was only in May, 1830, that the public opening took place. Worked on a series of inclined planes, partly by locomotives and partly with fixed engines, the Canterbury and Whitstaple

Railway was a sufficiently remarkable undertaking to attract public curiosity; still the event of its opening was disposed of briefly in the newspapers, as a thing of no more importance than the making of a few miles of ordinary road. It was, in fact, left to the Liverpool and Manchester line, which was opened four months later (15th September) to arouse the press and the public to the fact that there was in existence a process for conveying passengers and goods along the surface of the earth immensely superior to anything known in the world's history. The use of locomotive engines both for passengers and goods was a process fully established. Two lines of railway in England and two in Scotland were daily proving its enormous value, yet the reports of the inauguration of the railway connecting the great port of Liverpool with the manufacturing centres of Lancashire—perhaps not less by the success of the proceedings than by the tragic death of a popular statesman by which they were saddened—were the first to open the eyes of the nation and the world to the fact that the ....vent of the iron horse was about to revolutionise not only travelling but trade, and to bring to light a new power which, whether for peace or war, was to distance and displace all existing methods of conveyance, and bring about a new social era.

Application was made to Parliament for leave to lay down a railway from Liverpool to Manchester—a work then become indispensable to those two increasing and important towns. At that period, and for some time afterwards, canal boats, and slow, heavy road waggons were the only available means for the transport of heavy goods or bulky merchandise. The charge for conveyance from London to Yorkshire amounted frequently to £13 per ton, and even at this high cost the service was very imperfect. Beneficial as canals had proved, they were becoming inadequate to the growing requirements of trade. Besides the road there were two canals for the traffic between Liverpool and Manchester, the distance by the latter fifty-five miles, and the carriage of goods in some instances £2 per ton. Manchester was so entirely dependent on Liverpool that better accommodation became a necessity. Another canal could not be made, so a railway was projected; and the prospectus being issued in 1824, an Act was obtained, after failure in the session of 1825, in the year 1826.

It was the intention of its projectors to run the carriages upon it at a high rate of speed. To do this with horses was expensive; and to work it by steam-power, it was supposed that stationary engines would be required at short intervals along the road, to draw the trains by ropes from one station to another. While the necessity for the projected railway was admitted on all hands, the idea of its being worked by locomotives at a speed exceeding eight or nine miles an-hour was ridiculed. And when George Stephenson stated that he could make the locomotive travel at the rate of twenty miles an-hour, it was received with incredulity, and doubts were whispered as to his sanity. A reviewer in the *Quarterly* stated that nothing could be more palpably absurd than the prospect held out of locomotives travelling twice as fast as stage-coaches, and that people would as soon suffer themselves to be fired off upon one of Congreve's *ricochet* rockets as trust themselves to the mercy of a machine going at such a rate. When examined before a parliamentary committee, Stephenson's estimate of speed caused one member of the committee to remark that the engineer could only be fit for a lunatic asylum. The following case was put before Stephenson: "Suppose, now, one of those engines to be going along a railroad at the rate of nine or ten miles an-hour, and that a cow were to stray upon the line and get in the way of the engine, would not that, think you, be a very awkward circumstance?" "Yes," replied the witness, with a twinkle in his eye, "very awkward indeed—*for the coo!*" The honourable member did not proceed further with his cross-examination; and, says Dr. Smiles, "to use a railway phrase, he was *shunted*."

A premium of £500 was at length offered for the best engine, one that should not produce smoke, should draw three times its own weight for thirty miles, at the rate of ten miles an-hour, should be supported on springs, should not weigh more than six tons, and should not cost more than £550. At the time appointed, four locomotives were presented for trial, and the competition took place on the 6th of October, 1829, before many thousand spectators. One of the competing engines, the "Perseverance," made by Mr. Burstall, being found unable to move at more than five or six miles an-hour, was withdrawn. Another, called the "Novelty," made by Messrs. Braithwaite & Ericsson, was unable to complete the trial owing to the bellows for creating the blast having given way. A third, called the "Sanspareil," submitted by Mr. Hackworth, succeeded in drawing a load at the rate of fourteen miles an-hour; but at its eighth trip along the two-mile level which formed the course, the cold-water pump got wrong, and it could proceed no further. The "Rocket," made by George Stephenson, however, made an experimental trip of twelve miles, which was performed without accident in about fifty-three minutes.

Another day was fixed for the final trial of the competing engines; and on the morning of the 8th of October, the "Rocket" was again ready for the contest. "On that occasion," says Dr. Smiles, "the engine was taken to the extremity of the stage, the fire-box was filled with coke, the fire lighted, and the steam raised until it lifted the safety-valve, loaded to a pressure of 50 lbs. to the square inch. This proceeding occupied fifty-seven minutes. The engine then started on its journey, dragging after it about thirteen tons' weight in waggons, and made the first ten trips, backwards and forwards along the two miles of road, running the thirty-five miles including stoppages, in an hour and forty-eight minutes. The second ten trips were in like manner performed in two hours and three minutes. The maximum velocity attained during the trial trip was twenty-nine miles an-hour, or about three times the speed that one of the judges of the competition declared to be the limit of possibility! The average speed at which the whole of the journeys were performed was fifteen miles an-hour, and five beyond the rate specified in the conditions published by the company. The entire performance excited the greatest astonishment amongst the assembled spectators. The directors felt confident that their enterprise was now on the eve of success; and George Stephenson rejoiced to think that, in spite of all false prophets and fickle counsellors, his locomotive system was safe. When the 'Rocket,' having performed all the conditions of the contest, arrived at the 'grand stand' at the close of its day's successful run, Mr. Cropper, one of the directors favourable to the fixed-engine system, lifted up his hands and exclaimed, 'Now has George Stephenson at last delivered himself.' This interesting engine, the parent of the locomotives at present in use, is still to be seen in the Patent Museum at South Kensington."

The prize of £500 was at once awarded to the maker of the "Rocket." The engine was not only remarkable for its speed, but also for the contrivances by which that speed was attained. Most important among them was the introduction of tubes passing from end to end of the boiler—said to have been suggested by Mr. Booth, secretary to the company—by means of which so great an additional surface was exposed to the heat of the fire, that steam was generated much more rapidly, and a higher temperature maintained at a smaller expenditure of fuel than usual. The tubular boiler was indeed the grand fact of the experiment. Without tubes steam could never have been produced with the rapidity and heat essential to quick locomotion; and by burning coke instead of coal, the stipulated suppression of smoke was effected. The quantity of fuel consumed by the "Rocket" during the experiment was half-a-ton, the coke and water being carried in a tender attached to the engine.

One of George Stephenson's crowning achievements was the formation of the Manchester and Liverpool line; a project which, despite the sarcasms and incredulity with which it was assailed, succeeded beyond the engineer's most sanguine hopes.

"Mr. George Stephenson," says Dr. Smiles in his most interesting biography, "was no sooner appointed engineer than he removed his residence to Liverpool, and made arrangements to commence the works. He began with the impossible—to do that which the most distinguished engineers of the day had declared that 'no man in his senses would undertake to do'—namely, to make the road over Chat Moss! The drainage of the moss was commenced in June, 1826. It was indeed a most formidable undertaking; and it has been well observed that to carry a railway along, under, or over such material as the moss presented, could never have been contemplated by an ordinary mind. Michael Drayton supposed Chat Moss to have had its origin at the Deluge. Nothing more impassable could have been imagined than that dreary waste; and Mr. Giles only spoke the popular feeling of the day when he declared that no carriage could stand on it 'short of the bottom.' In this bog, singular to say, Mr. Roscoe, the accomplished historian of the *Medicis*, buried his fortune in the hopeless attempt to cultivate it. Nevertheless, farming operations had for some time been going on, and were extending along the verge of the moss; but the tilled ground, underneath which the bog extended, was so soft that the horses when ploughing were provided with flat-soled shoes to prevent their hoofs sinking deep into the soil.

"For weeks the stuff was poured in, and little or no progress seemed to have been made. The directors of the railway became alarmed, and they feared that the evil prognostications of the eminent civil engineers were now about to be realised.

"Mr. Stephenson was asked for his opinion; and his invariable answer was, 'We must persevere." And so he went on; but still the insatiable bog gaped for more material, which was emptied in truck-load after truck-load without any apparent effect. Then a special meeting of the board was summoned, and it was held upon the spot, to determine whether the work should be proceeded with or *abandoned*. Mr. Stephenson himself afterwards described the transaction at a public dinner given at Birmingham on the 23rd of December, 1837, on the occasion of a piece of plate being presented to his son, the engineer of the London and Birmingham Railway. He related the anecdote, he said, for the purpose of impressing upon the minds of all who heard him the necessity of perseverance.

"'After working for weeks and weeks,' said he, 'in filling in materials to form the road, there did not yet appear to be the least sign of our being able to raise the solid embankment one single inch; in short, we went on filling in without the slightest apparent effect. Even my assistants began to feel uneasy, and to doubt of the success of the scheme. The directors, too, spoke of it as a hopeless task, and at length they became seriously alarmed; so much so, indeed, that a board meeting was held on Chat Moss to decide whether I should proceed any further. They had previously taken the opinion of other engineers, who reported unfavourably. There was no help for it, however, but to go on. An immense outlay had been incurred; and great loss would have been occasioned had the scheme been then abandoned and the line taken by another route. So the directors were *compelled* to allow me to go on with my plans, of the ultimate success of which I myself never for one moment doubted. Determined, therefore, to persevere as before, I ordered the work to be carried on vigorously; and to the surprise of every one connected with the undertaking, in six months from the day on which the board had held its special meeting on the moss, a locomotive engine and carriage passed over the very spot with a party of the directors' friends on their way to dine at Manchester.'"

The works were so far advanced that the line was expected to be ready for opening on 1st January, 1830. A desire to secure greater prominence to the event postponed the opening to a later date, and it was not till the 15th of September that the inauguration took place.

After the opening of the Liverpool and Manchester line, Stephenson took up his residence near Ashby-de-la-Zouch, in Leicestershire, where he resided for some years. In company with two Liverpool friends, he purchased an estate in the neighbourhood of Ashby, called Snibston, where he sank several shafts; and after overcoming no common difficulties, by means of the Leicester and Swannington Railway, the people of Leicester were enabled to purchase coals at 8s. a-ton. This

Dr. Smiles remarks, must have had the effect of saving about £40,000 per annum to the inhabitants. His correspondence had now increased so much that he was obliged to engage a private secretary. In the year 1835, during his busy season, he dictated no fewer than thirty-seven letters daily. At another time, he dictated letters and reports for twelve hours at a time, until his secretary was like to drop from his chair with fatigue, and was obliged to ask for a suspension from labour. These years ending 1837 are said by his biographer to have been the busiest of his life; amongst other duties, he was engaged in the survey as principal engineer of the North Midland Railway from Derby to Leeds, the York and North Midland from Normanton to York, the Manchester and Leeds, the Birmingham and Derby, and the Sheffield and Rotherham Railways. In 1841 he said that there was hardly a railway in England which he had not had to do with. In the survey of a proposed line in Scotland between Edinburgh and Newcastle, by the vale of the Gala (now occupied by the Waverley route), and by Carter Fell on the Cheviots, as against a coast route by way of Berwick-on-Tweed, his report was strongly in favour of the latter. The reasons were apparent—the railway would follow that low coastline, possessing levels of a favourable nature, and being near the coast the snow would not lie so long on the line. The rush for railway Acts, as Dr. Smiles tells us, was now extraordinary. In the year 1836, thirty-four Bills passed the Legislature, authorising the making of 994 miles of new railway, the cost being estimated at £17,595,000. During 1837 no less than 118 notices of new railway Bills were given in. Of these seventy-nine were introduced to Parliament, forty-two Acts were obtained, fourteen companies were incorporated and authorised to construct 464 miles of railway at a cost of

£8,087,000. From 1838 till 1844 the desire for new railway works was moderated, and in 1845 another tide of railway speculation set in. Powers were granted by Parliament during that year to construct 2883 miles of new railways in Britain, and next year the power was conceded for still larger undertakings.

Speaking at Blackburn in 1840, Stephenson intimated his intention of retiring from the more arduous portion of his duties, which he did by resigning his charge as chief engineer on several of the railways with which he was connected. He had removed in 1838 to Tapton House, near Chesterfield, and he was able to enjoy the grounds, which he greatly improved. His name and presence were common at this time with the Mechanics' Institutes of the Midland Counties. In 1844 he was appointed engineer to the Whitehaven and Maryport Railway, and at about the same time was elected chairman of the Yarmouth and Norwich Railway. The Newcastle and Darlington line was completed and publicly opened in 1844. By this line direct communication was gained with London. Mr. Stephenson, Mr. Hudson, and a distinguished party, travelled from London to Newcastle in nine hours. At the banquet in the evening, the honourable Mr. Liddell, M.P., occupied the chair, and paid a high compliment to George Stephenson and his son Robert. "Mr. Stephenson," he said, "might truly be looked upon as the great pacificator of the age. And yet a few years ago he was but a working engineman at a colliery. But he was a man not only of talent but of genius. Happily, also, he was a man of industry and of character. He constructed the first successful engine that travelled by its own spontaneous power over an iron railroad, and on such a road, and by such an engine, a communication had now been established between London and Newcastle. The author of this system of travelling had lived long enough for his fame but not long enough for his country. He had reared to himself a monument more durable than brass or marble, and based it on a foundation whereon it would rest unshaken by the storms of time." Stephenson replied as follows: "As the honourable member," said he, "has referred to the engineering efforts of my early days, it may not be amiss if I say a few words to you on that subject, more especially for the encouragement of my younger friends. Mr. Liddell has told you that in my early days I worked at an engine at a coal-pit. I had then to work early and late, and my employment was a most laborious one. For about twenty years I had often to rise to my labour at one and two o'clock in the morning, and worked till late at night. Time rolled on, and I had the happiness to make some improvements in engine-work. The company will be gratified when I tell them that the first locomotive that I made was at Killingworth Colliery. The owners were pleased with what I had done in the collieries, and I then proposed to make an engine to work upon the smooth rails. It was with Lord Ravensworth's money that my first locomotive was built. Yes, Lord Ravensworth and his partners were the first gentlemen to entrust me with money to make a locomotive. This was more than thirty years ago, and we first called it 'My Lord.' I then stated to some of my friends now living that those high velocities with which we are now so familiar would sooner or later be attained, and that there was no limit to the speed of such an engine, provided the works could be made to stand; but nobody would believe me at that time. The engines could not perform the high velocities now reached when they were first invented, but, by their superior construction, an immense speed is now capable of being obtained. In what has been done under my management, the merit is only in part my own. Throughout I have been most ably seconded and assisted by my son. In the early period of my career, and when he was a little boy, I felt how deficient I was in education, and made up my mind that I would put him to a good school. I determined that he should have as liberal a training as I could afford to give him. I was, however, a poor man; and how do you think I managed? I betook myself to mending my neighbours' clocks and watches at night after my daily labour was done. By this means I saved money, which I put by; and in course of time I was thus enabled to give my son a good education. While quite a boy he assisted me and became a companion to me. He got an appointment as under-viewer at Killingworth, and at nights when we came home we worked together at our engineering. I got leave from my employers to go from Killingworth to lay down a railway at Hetton, and next to Darlington for a like purpose; and I finished both railways. After that I went to Liverpool to plan a line to Manchester. The directors of that undertaking thought ten miles an-hour would be a maximum speed for the locomotive engine, and I pledged myself to attain that speed. I said I had no doubt the locomotive might be made to go much faster, but we had better be moderate at the beginning. The directors said that I was quite right; for if, when they went to Parliament, I talked of going at a greater rate than ten miles an-hour, I should put a cross on the concern. It was not an easy task for me to keep the engine down to ten miles an-hour, but it must be done; and I did my best. I had to place myself in the most unpleasant of all positions—the witness-box of a parliamentary committee. I was not long in it, I assure you, before I began to wish for a hole to creep out at

I could not find words to satisfy the committee or myself, or even to make them understand my meaning. Some said, 'He's a foreigner.' 'No,' others replied, 'he's mad.' But I put up with every rebuff, and went on with my plans, determined not to be put down. Assistance gradually increased; great improvements were made in the locomotive, until to-day a train which started from London in the morning has brought me in the afternoon to my native soil, and enabled me to meet again many faces with which I am familiar, and which I am exceedingly pleased to see once more."

When the question of the use of atmospheric railways *versus* the locomotive came before Parliament in 1845, the locomotive triumphed. "The king of railway structures," the high-level bridge over the Tyne, was first promulgated by Mr. R. W. Brandling in 1841. The designs for the bridge were Mr. Robert Stephenson's, and the name of George Stephenson appeared in the committee of management. In 1835 George Stephenson and his son had been consulted as to the establishment of an efficient railway system throughout Belgium. During a visit to Belgium the king appointed him a Knight of the Order of Leopold. From the opening of the first lines in 1835 until 1844, about six and a-half millions sterling had been laid out in railways there. The Belgian lines being executed as an entire system by the State, their railway system is said by Dr. Smiles to have averaged in cost less than one-half that of England. Stephenson was present at the public opening of the line from Brussels to Ghent, and on the day following had the honour of dining with the king and queen at their own table at Laaken. The engineers of Brussels entertained him to a magnificent banquet. A model of the " Rocket" was placed upon the centre table at dinner, under a triumphal arch. The king had a private interview with him, when he is said to have been easy and self-possessed while explaining to him the structure of the Belgian coalfields, with the rise and progress of trade and manufactures, which were so closely dependent for their existence on these coalfields. In 1845 he visited Spain, in order to report on "The Royal North of Spain Railway," but his report was so unfavourable that the scheme was abandoned.

Towards the close of his days, George Stephenson lived the life of a country gentleman, and began to take a greater interest than he had ever previously time or opportunity to do in horticultural pursuits. He built melon-houses, pineries, and vineries, and workmen were continually busied in his garden. He also carried on farming operations to some extent. His old love for birds and other animals returned, and he soon knew every bird's nest in the neighbourhood. He read few books, loving to gather most of his information from intelligent conversation, and the most of his letters were dictated. His simple tastes in eating and drinking continued to the last, and he could still enjoy a bowl of "crowdie," a dish made with oatmeal and boiling water—a reminiscence of his pitman days. He despised foppery of all kinds, and one day when a young man, desirous to be an engineer, came to him for advice, flourishing a gold-headed cane, he said, "Put by that stick, my man, and then I will speak to you." To another affectedly-dressed young man he one day said, "You will, I hope Mr. ——, excuse me; I am a plain-spoken person, and am sorry to see a nice-looking and rather clever young man like you disfigured with that fine-patterned waistcoat, and all these chains and fang-dangs. If I, sir, had bothered my head with such things when at your age, I would not have been where I am now." Occasionally he would visit old scenes at Newcastle, but generally avoided, if possible, what was called fine company; he also refused Sir Robert Peel's offers of knighthood. Projectors of various kinds would seek his advice on subjects connected with mechanical engineering. During the last year of his life Mr. Stephenson met Ralph Waldo Emerson, the great American thinker and writer. Emerson afterwards remarked "that it was worth crossing the Atlantic to have seen Stephenson alone; he had such native force of character and vigour of intellect." Universally respected, and simple-minded and upright to the last, George Stephenson passed away on the 12th August, 1848, in his sixty-seventh year.

Dr. Smiles has drawn the following word-portrait of the great engineer :—" His fair, clear countenance was ruddy, and seemingly glowed with health. The forehead was large and high, projecting over the eyes; and there was that massive breadth across the lower part, which is usually observed in men of eminent constructive skill. The mouth was firmly marked; and shrewdness and humour lurked there as well as in the keen grey eye. His frame was compact, well-knit, and rather spare. His hair became grey at an early age, and towards the close of his life it was of a pure silky whiteness. He dressed neatly in black, wearing a white neck-cloth; and his face, his person, and his deportment at once arrested attention, and marked the gentleman." A hall devoted to scientific, literary, and educational purposes, and costing £13,000, has been erected to his memory at Chesterfield, where he once resided for a time.

His distinguished son Robert only survived him eleven years. His death took place in 1859, when in his fifty-sixth year, and he received the honour of a public funeral, and was interred in Westminster Abbey. He designed the high-level bridge across the Tyne at Newcastle, and with Sir William Fairbairn of Manchester he divides the honour of the Conway and Britannia tubular bridges in North Wales. The honour of the invention of the tubular system of bridge-building rests with Fairbairn. The tubular bridge across the St. Lawrence at Montreal was also Stephenson's design.

SIR JAMES Y. SIMPSON.

BATHGATE, a town of over 5000 inhabitants, lies in a district rich in coal, shale, ironstone, and limestone, about eighteen miles west from Edinburgh. Looking southwards from the slope of the hill overlooking the town, the eye meets the line of the Pentlands, intersected with deep valleys and ravines; the country between is dotted with oil-works, which at night look like a village on fire. Eastward Arthur's Seat and Edinburgh Castle loom largely through the haze of distance; westward is an undulating agricultural tract; northwards the eye meets the blue line of the Firth of Forth, extending upwards towards Stirling. The town probably did not number over 3000 inhabitants when James Young Simpson was born there, 7th June, 1811. His immediate ancestry on both father and mother's side came of a good farmer's stock. Further back, on his mother's side, he was allied with the gentle blood of Scotland; on the father's side with a race of vigorous limb, strong will, and great shrewdness and industry. His father's name was David Simpson, and his mother's Mary Jarvey. At the time of the birth of the seventh son and eighth child, the fortunes of David Simpson were at a very low ebb. The drawings in the baker's shop, on the day on which James Y. Simpson was born, amounted to 8s. 3d. Mrs. Simpson, a woman of energy and tact, discovering this state of affairs, turned her attention to the details of the business, which afterwards continued fairly prosperous.

Mary Simpson was one of the best of mothers, always displaying much force of character, along with a quiet, loving disposition. She died when James was nine years of age, but the memory of her prayers remained with him through life. The cares of the household then fell upon his only sister Mary. He was sent to school when four years of age. His school tasks were easy work for him, and his love of knowledge and of a good bone of fact were insatiable; so much so, that on overhearing some one say that "the Bible and Shakespeare are the best books in the world," he remarked, "The Bible and Shakespeare, and Oliver & Boyd's Almanac! At least I know the Almanac would have been the greatest prize for me when a boy." At home he was gentle and obliging, and made himself useful in the shop, and in delivering bread around the neighbourhood. He never felt the straits of pinching poverty which so nearly threatened his elder brothers, the business having prospered from the date of his birth. During one of his earliest visits to Edinburgh he made his way direct to Greyfriars Churchyard, where he copied an inscription on one of the tombstones.

With an early longing for student life he entered Edinburgh University at the age of fourteen, attending the junior Greek and Humanity classes in session 1825-26, under Professors Dunbar and Pillans. In 1827-28 he enrolled as a student of medicine, and during the same session attended the classes of Natural Philosophy, Moral Philosophy, and the third Greek class. When entering on his second session he was fortunate enough to secure a bursary of the value of £10, tenable for three years. He had joined an old friend, John Reid, who then lodged with Dr. Macarthur, in No. 1 Adam Street. One of the first books he bought on coming to Edinburgh was "The Economy of Human Life," for which he paid ninepence. He was strictly economical, and kept an exact note of his expenses, which at the end of the session he submitted to the family. The rent of his room was not more than three shillings a-week. His little cash-book contained such entries as the following :— "Vegetables and 'Byron's Beauties';" "Finnen Hadies, 2d.; and Bones of the Leg, £1, 1s.;" "Subject, £2; Spoon, 6d.; and Bread and Tart, one shilling and eightpence; Fur Cap, 14sh.; Mary's Tippet, 2sh. and 6d.; Duncan's 'Therapeutics,' 9d.;" "Snuff, 1½d.; and a book on 'Early Rising,' 9½d." One of his father's letters to him, written in 1826, ran as follows :—

"My Dear Son,—I am glad to hear by John Pearson that you are well. I intended to be in Edinburgh this month, but I find it is out of my power. Be so good as write me what money you will take to bring you out. James, I am now turning old, and wearing awa' like the snaw among the thaw. I have had a weary winter, but will be glad to see you at Bathgate." David Simpson died in 1839.

He attended closely, and benefited largely from Professor Liston's surgery classes. It was while attending Liston's surgery classes, and witnessing the operations, "that he first began to grope after means for the alleviation of pain when the patient was in the hands of the operator. After seeing the terrible agony of a Highland woman under amputation of the breast, he left the class-room, and went straight to the Parliament House to seek work as a writer's clerk." So the question, "Can anything be done to make operations less painful?"

became a pressing one with him. He passed with ease and credit in the examination for his degree, and became a member of the Royal College of Surgeons before he was nineteen years of age. As he could not take his degree as a physician until he was twenty-one, he returned for a time to Bathgate. He entered college again in 1831, and became first assistant to Dr. Gairdner in dispensary work, to whom he gave the utmost satisfaction. In 1832 he received the degree of Doctor of Medicine. Referring to this period of his life, long afterwards, he said : " 'Tis fully forty years since I came first to Edinburgh, and entered its university as a very, very young and very solitary, very poor, and almost friendless student. . . . Nor was my original ambition in any way very great. After obtaining my surgical diploma, I became a candidate for a situation in the West of Scotland, for the attainment of which I fancied I possessed some casual local interest. The situation was surgeon to the small village of Inverkip, on the Clyde. When not selected I felt perhaps a deeper amount of chagrin and disappointment than I have ever experienced since that date. If chosen, I would probably have been working there as a village doctor still. But like many other men I have, in relation to my whole fate in life, found strong reason to recognise the mighty fact, that assuredly—

> 'There's a Divinity that shapes our ends,
> Rough hew them how we will.'

Yes, in the language of the French proverb, 'Man proposes, but God disposes.'"

When his literary studies at the university were finished, he threw himself, heart and soul, into his medical studies. Dr. John Thomson, an eminent physician, saw in young Simpson a possible assistant, and engaged him at the modest salary of

£50 a-year. This sum he made to suffice for all his wants. "Professor Thomson," he wrote, "engaged me as his assistant, and hence, in brief, I came to settle down a citizen of Edinburgh, and fight amongst you a hard and uphill battle of life for bread, and name, and fame." Between the years 1831-36, at Dr. Thomson's request, he turned his attention to the study of obstetric medicine. Already he had shown marked powers of concentration of mind, and in commendation of work and diligence he once said: "Sir Isaac Newton, whose gigantic genius and intellectual strength have fixed upon him the admiration of his race, modestly averred that his mental superiority, if any, consisted, in his own opinion, only of unusual powers of patient thought and industry. The unparalleled greatness in the results of his thoughts was owing, according to his own interpretation, merely to his habit of unparalleled endurance and assiduity in the exercise of thinking." When he was cautioned by his sister Mary in 1834 for working too hard, he said: "Well, I'm sure, it's just to please you all." In 1833 he became a member of the Royal Medical Society of Edinburgh, and in 1835, in company with Dr. Douglas Maclagan, and by the ever-ready liberality of his brothers, Alexander and John, he was enabled to visit the chief schools of medicine, and the hospitals of England and France. In session 1835-36 he was elected senior president of the Royal Medical Society. His appearance, when presiding at one of the meetings of the society, has been thus described:—"As we entered the room his head was bent down, to enable him, in his elevated position to converse with some one on the floor of the apartment, and little was seen but a mass of long, tangled hair, partially concealing what appeared to be a head of very large size. He raised his head, and his countenance at once impressed us.

A poet has since described him as one of 'leonic aspect. Not such do we remember him. A pale, large, rather flattish face, massive, brent brows, from under which shone eyes, now piercing as it were to your inmost soul, now melting into almost feminine tenderness, a coarsish nose, with dilated nostrils, a finely-chiselled mouth, which seemed the most expressive feature of the face, and capable of being made at will the exponent of every passion and emotion. Who could describe that smile? When even the sun has tried it he has failed; and yet who can recall those features and not realise it as it played round the delicate lines of the upper lip, where firmness was strangely blended with other and apparently opposing qualities?" His general deportment has thus been summed up : "He could be considerate in a sick room, genial at a feast, joyous at a festival, capable of discourse with many minds, large-souled, not to be shrivelled up into any one form, fashion, or temperament." When rapidly climbing the ladder of life, and when he was daily gaining an increased practice, he still kept himself interested in all his old schoolmates and Bathgate friends.

We now come to a turning-point in his career. Professor Hamilton resigned the Midwifery chair in the university in 1839, and Dr. Simpson announced himself as a candidate. By hard work and a stiff canvass amongst the members of the town council, who were the patrons, he was elected by a majority of *one*. Partly because his celibacy had been an objection to his election, and now that his position fairly warranted such a change, he was married, on the 26th December, 1839, to Miss Jessie Grindlay, daughter of Mr Grindlay, shipowner, Liverpool. The honeymoon was protracted until after the election, and his Liverpool letter, telling of his success, ran thus :—

"My Dear Mother,—Jessie's honeymoon and mine is to begin to-morrow. I was elected professor to-day by a MAJORITY OF ONE. Hurrah !!!—Your ever affectionate Son,

"J. Y. SIMPSON."

The congratulations of his sister Mary, who was just leaving for Van Dieman's Land with her husband, were very true, and tender, and sisterly. "My dear, dear, and fortunate Brother," she wrote, "I have taken up my pen to wish you joy, joy; but I feel I am scarcely able to write. I never believed till now that excess of joy was worse to bear than excess of grief. I cannot describe how, but I certainly feel as I never did all my life. I hope we will still be here to-morrow to learn all the particulars of this happy event. My dear, dear James, may God Himself bless you, and prosper you in all your ways.—Your sincerely affectionate Sister, MARY PEARSON." And long afterwards she again wrote to his wife: "I am delighted with your description of your dear little Maggie. All that you write of James, *my* James (for I have both a mother's and a sister's love for him), is just what I expected. I knew he would be as kind a husband as a brother. It is with feelings of proud joy I hear of his unbounded success. Long, long may he be spared to be useful to others, and a joy to us all." Professor Duns, his biographer, truly remarks that "Mary had guided him from childhood with deep and watchful love. In her care and keeping he had often felt 'as one whom his mother comforteth.' Her praise had frequently been to him an excitement and spur to exertion. The care with which he preserved her note from 'Ramsgate Harbour,' shows how much calm content it had given him— linking, as it forcibly did, this triumph in a great contest with the memory of all the love and encouragement that he had received from her and his brothers, but chiefly from her and his beloved "Sandy.'"

The members of the medical profession, it has been remarked, are often peculiarly conservative regarding new discoveries or new methods of work. Simpson was entirely free from this, indeed, "in the advocacy or defence of new methods of treatment, or of new remedies, he seldom took into account the prejudices, or even the honest convictions, of others." In his first session as professor, he met with some sharp antagonism, which he repelled as sharply. Regarding the way in which he did his professorial work, Professor Duns writes: "His genial bearing among the students, the earnest way in which he did the work of the class, the forcible and lucid style of his prelections, his breadth of view as a public teacher, the pleasant talk and sallies of quiet humour with which he often relieved the dry exposition of methods of research or the didactic statement of principles, the abundance and freshness of his illustrative facts, and his happy art of laying fields of thought outside of his profession under contribution, to give force and clearness to the special topics under review, all conduced to make him a favourite of the students generally, and to win the admiration, respect, and love of those who were foremost among them in mental power and accomplishments."

An address which Professor Simpson delivered to medical students contains the following high ideal of his profession:—

"The profession is," he said, "in many respects, the most important secular profession which a man can follow. Its importance depends on the priceless value of the objects of the physician's care and study, viz., the guardianship of the health and lives of our brother men, and the defence of the human body and human mind against the attacks and effects of disease.

Other pursuits become insignificant in their objects when placed in comparison with this. The agriculturist bestows all his professional care and study in the rearing of crops and cattle; the merchant spends his energies and attention on his goods and his commissions; the engineer, upon his iron wheels and rails; the sailor, upon his ships and freights; the banker, upon his bills and his bonds; and the manufacturer, upon his spindles and their products. But what, after all, are machinery and merchandise, shares and stocks, consols and prices current, or the rates of cargoes and cattle, of corns and cottons, in comparison with the inestimable value and importance of the health and the very lives of those fellowmen who everywhere move, and breathe, and speak, and act around us? What are any, or what are all, of these objects when contrasted with the most precious and valued gift of God to earth—human life? And what would not the greatest and most successful followers of such varied callings give out of their professional stores for the restoration of health and for the prolongation of life, if the first were once lost to them, or if the other were merely menaced by the dreaded and blighted finger of disease? . . . Nature has happily ordained it as one of the great laws on which she has founded our moral happiness, that the performance of love and kindness to others should be a genuine and never-failing source of pleasure to our own hearts. It is thus strictly as well as poetically true :

'That seeking others' good, we find our own.'

"The exercise of the profession is, when followed out in its proper spirit, a continued realisation of active beneficence, and, in this view, a continued source of moral satisfaction and happiness to the generous heart. The objects and powers of your art are alike great and elevated. Your aim is, as far as possible, to alleviate human suffering and lengthen out human existence. Your ambition is to gladden as well as to prolong the course of human life, by warding off disease as the greatest of mortal evils, and restoring health, and even, at times, reason itself, as the greatest of mortal blessings. If I may borrow the beautiful language of the author of "The Village":

> "'Glorious your aim—to ease the labouring heart,
> To war with death, and stop his flying dart;
> To trace the source whence the fierce contest grew,
> And life's short lease on easier terms renew;
> To calm the phrensy of the burning brain,
> And heal the tortures of imploring pain;
> Or, when more powerful ills all efforts brave,
> To ease the victim no device can save,
> And smooth the stormy passage to the grave.'

I repeat it—if you follow these, the noble objects of your profession, in a proper spirit of love and kindness to your race, the pure light of benevolence will shed around the path of your toils and labours the brightness and beauty that will cheer you onwards, and keep your steps from being weary in well-doing; while, if you practise the art that you profess with a cold-hearted view to its results merely as a matter of lucre and trade, your course will be as dark and miserable as that low and grovelling love that dictates it."

He was now a prosperous man, and in the first three months of 1844 he received £1000 as fees. At the close of 1845, his practice was still increasing amongst the nobility and aristocracy. He bought a house at No. 52 Queen Street, where, even at this early date, patients sought him in such numbers, that his limited time and other engagements would not permit him to overtake half of them. Country cases received his special attention. His correspondence increased, and he received frequent visits from strangers—English, Continental, and American—who all made demands upon his time. Those who were still disaffected towards him spread the report that his patients were neglected. One man of wealth enclosed him £10 where £100 might have been expected. He received an urgent and insulting note from the friends of a lady in the country for inattention. It turned out that the "very urgent matter" was whether three leeches should be applied to the hip-joint, as recommended by him, or if two only should be applied, as recommended by the country practitioner. In January, 1847, he received the honour of appointment as one of her Majesty's physicians for Scotland, regarding which the Queen, in a private letter to the Duchess of Sutherland, had said, "his high character and abilities make him very fit for."

When a young man, as we have already noticed, he had been much affected by the terrible agony endured by a Highland woman who had been undergoing a surgical operation. Aware of the fact that, by inhaling sulphuric ether, the patient was rendered insensible to pain, during the summer of 1847 he could think of naught else. He contributed a paper to the *Monthly Journal of Medical Science*, entitled, "Notes on the Inhalation of Sulphuric Ether in the Practice of Midwifery." These notes were separately printed and widely distributed amongst professional men at home and abroad. Turning his attention to other drugs, he became convinced that chloroform was much superior to ether for the purpose he had in view, and in March, 1847, he communicated a paper on its use to the Medico-Chirurgical Society of Edinburgh. From these notes we learn that "chloroform was first discovered and described at nearly the same time by Soubeiran (1831) and Liebig (1832); its composition was first accurately ascertained by the distinguished

French chemist, Dumas, in 1835. . . . It is a dense, limpid, colourless liquid, readily evaporating, and possessing an agreeable, fragrant, fruit-like odour, and a saccharine pleasant taste." Cases were cited wherein its use had been successful. These investigations were afterwards published in a pamphlet form, of which four thousand copies were sold in a few days. Believing that he could only at first truly test the value of his discovery by personal experience, he, as a matter of course, tried its effects on himself. "On the first occasion," he writes, "on which I detected the anæsthetic effects of chloroform, the scene was an odd one. I had had the chloroform beside me for several days, but it seemed so unlikely a liquid to produce results of any kind, that it was laid aside, and on searching for another object among some loose paper, after coming home very late one night, my hand chanced to fall upon it, and I poured some of the fluid into tumblers before my assistants, Dr. Keith and Dr. Duncan, and myself. Before sitting down to supper, we all inhaled the fluid, and were all 'under the mahogany' in a trice, to my wife's consternation and alarm. In pursuing the inquiry thus rashly, perhaps, begun, I became every day more and more convinced of the superior anæsthetic effects of chloroform, as compared with ether." Several attempts were made both in England and America to rob him of the honour of his discovery, which led to a prolonged controversy. His butler, Clarke, who had a high opinion of the properties of "chlory," as he called chloroform, found him lying in his room, apparently unconscious, and suffering from the effects of a recent experiment. To some friends who were watching him, with a good deal of alarm, the butler remarked, "He'll kill himsel' yet wi' thae experiments; an' he's a big fule, for they'll never find onything better nor chlory." On another occasion, he rushed into the dining-room, saying, "For God's sake, sir, come doun; I've pushioned the cook." When Dr. Simpson, with some others, got downstairs, they found the patient lying on the floor snoring heavily. This incident created a hearty laugh. The butler had handed the cook the remains of an effervescing drink, prepared with chloric ether, calling it champagne. She drank it with the above result. Dr. Simpson had also to appear in print regarding what was termed the religious side of the use of chloroform. Many quoted the primary curse in Genesis as a reason why chloroform should not be used in midwifery. These would-be religious objections Dr. Simpson successfully disposed of, and, after more than twenty years had elapsed since its discovery, he found the best answer to all arguments of this kind was to point to the steady increase in its production, and to its ever-extending use.

About this time a medical officer of the Indian Army wrote thus to the *Bombay Telegraph and Courier:* "Decidedly the most wonderful man of his age—I mean of the age in which he lives—is Simpson of Edinburgh. In him are realised John Bell's four ideals of the perfect Esculapius—the brain of an *Apollo,* the eye of an *Eagle,* the heart of a *Lion,* and the hand of a *Lady.* Nothing baffles his intellect; nothing escapes his penetrating glance; he sticks at nothing and *he bungles nothing.* If his practice be worth a rupee per annum, it is worth £10,000 —twice as much as Dr. Hamilton ever realised, and nearly twice the amount of the late Abercrombie's practice. From all parts, not of Britain only, but of Europe, do ladies rush to see, consult, and fee the man. He has spread joy through many a rich man's house by enabling his wife to present him with a living child, a feat which none but Simpson ever dared to enable her to do. To watch of a morning with his poor patients (them only of course was I permitted to see) is a treat. In comes a woman with a fibrous tumour, which fifty other practitioners have called by fifty other names. One minute suffices for his diagnosis; another sees her in a state of insensibility, and in less than a third, two long needles are thrust inches deep into the tumour, and a galvanic battery is at work discussing it. 'Leave her alone quietly,' says Simpson, 'she'll take care of herself—no fear.' One up, another down, is the order of the day. What other men would speculate as to the propriety of for hours, Simpson does in a minute or two. He is bold, but not reckless; ever ready, but never harsh. He is prepared for every contingency, and meets it on the instant. Everything seems to prosper in his hands. As to ether and chloroform, they seem like invisible intelligences, doomed to obey his bidding—familiars who do his work because they must never venture to produce effects one iota greater or less than he desires. While other men measure out the liquids, fumble about and make a fuss, Simpson, in what an Irishman would call the most promiscuous manner possible, does the job in a minute or two. He is indeed a wonderful man."

The account given by Dr. Channing, of Boston, U.S., of a visit to Queen Street is equally interesting: "At half-past one his consultations at home begin, and last till nearly or quite six. When he began this system of home clinics, for such they strictly are, his house was filled at all hours, so that it was impossible to keep any order. People would come at seven A.M. in order to be first. They would get breakfast at six, or earlier, and disturb their own families much. To prevent this, he fixed the hour at half-past one to half-past five. The patients of the two divisions are in different but equally large rooms. They draw lots for priority, have tickets, and come in as called, and so the most perfect order prevails. Everybody knows what are Professor S.'s hours, and everybody observes them. He has an assistant who writes prescriptions to his dictation, directions, letters, &c., and also attends to cases. He examines cases daily when there is occasion to do so. From long experience and constant observation, the habit of recording cases, and of distinguishing them with all the accuracy in his power, he is able to arrive at conclusions in the cases before him in a very short time, or to make his diagnosis. I see most, or many of his cases, examine them after him, and I have again and again been struck, in new ones, how true is his diagnosis. He proceeds at once to the treatment. If an operation is to be, he does it at once. Applications of remedies are made and prescriptions given, with directions, and the patient is desired to call in a week, fortnight, in two days, &c., as circumstances may indicate. At times the case is written down from the answers of patients to questions. This is always the case if it be a new case, or it is probable that changes may be required in treatment, or the effects of treatment noted. Some notion may be got of this portion of Professor S.'s indoor or home professional life. He goes through this great labour quietly and methodically, and with as gentle, kind, cheerful spirit as man ever manifested. The moral character of the daily service in disease is quite as striking as the professional. The moral presides over the whole, and renders it one of the most interesting matters for observation that can occur. I have been utterly surprised at its executive patience, its efficient activity. Here are the poor and the rich together, with no other distinctions than such as will best accommodate both. And I can say, from a long and wide observation, that there is no difference in their treatment. The great fact of each in Professor S.'s regard is the fact that disease exists, which it is the physician's business to investigate and try to remove. He knows what is the prospect of success or of failure, and makes his prognosis accordingly. But, even when the worst is announced, it is not spoken of as utterly hopeless, and something is done, all is done, for present comfort, when nothing may be done for cure. I am surprised again at the varieties of disease which congregate at No. 52, and of the number of which is presented in each kind. It is this which gives character to the whole, and makes these clinics the very best schools. I have been every day a pupil here. I have every day learned much; yes, a great deal, which will aid me in all my professional, yes, my moral life. I had designed to visit Ireland, but so few days remained to me that I was sure the visit could amount to nothing important, and I concluded to remain at Professor Simpson's house, in the midst of his home practice, and to visit with him abroad such patients as he could show me. Wherever we went the professor was received with the same bright welcome, the same cheerful face, and I thought this made the beauty of his professional life. One was glad to see him so soon again. Another had been waiting with such patience as could be commanded for a visit. But with all was the appearance and the consciousness that something good was to come from the call. He had time for everything. He took his seat, and with his 'Come along now, how are you? how have you been?' &c., was always answered to satisfy perfectly the various objects in view. There were directions in his questions, or directions to his patients; but it was so quiet, so easy, that, though time was pressing on new engagements, it seemed that the present one only occupied his mind. There was persuasion with command, or demand in such proportions that the patient was only anxious to do the very best for himself, or for herself, and for the doctor. In this way or by this manner, which seems no manner at all, Professor S. is able to do a great deal in a short time. His coachman understands by a hint where he is to go, and goes rapidly through his various service. As we pass along some object of interest is at hand—the Botanic Garden, a ruin, a hill, a beautiful prospect. He pulls the string, opens the door which lets down the step, and 'Come away,' tells you there is something for you to see, something to please you, and there is time enough to see it. 'I visit here, and for ten minutes I will leave you; go down there and you will find something.' Off he goes to his patient, and off I go to see what he has indicated. The Professor is well made for despatch. He is short, stout, with small feet, and his step is short and very quick. He is of excellent age for vigour—about thirty-nine—and 'goes ahead' of all walkers. I have almost to run somewhat, not to lose him. Let me finish his picture. You have his length, but not his full length. His head is large, covered with a profusion of black hair which obeys its instincts, and more strikingly so when he thrusts his very small hands into and all over it. His forehead is of good height, but the hairs grow low upon it; and to me this is the most becoming manner of its growth, and the antique, the Apollo, the Clyte, &c., support my taste. His face is broad, of fair length, and its expression just such as such mind and heart always produce. His eyes are singularly loquacious, and always begin to talk before he utters a word. His knowledge is more various than I have before met with. Nothing escapes him. Science and literature are his pleasures. Archæology is a favourite pursuit; and his friends frequently send him books and specimens, which help his studies. I never saw so many presents. I went up last night late. 'I must make some visits,' said he, 'say at eleven.' Off drove his coach. This morning, before anybody else was up, I went below for my spectacles. On the sideboard was a basket of fine peaches, 'which was not so before.' In the morning bouquets came in. I could fill pages with a list of such offerings as are daily poured in. He has game at every meal. 'Our friends,' said one, 'keep us supplied with game.' His family pass the summer in a very pleasant place a few miles from the city, but his house affairs go on by themselves, very much as of themselves, and knew how, and are all in perfect order. Said he to me when he carried me bodily from my hotel, 'I am a bachelor —no woman; but come away, you shall have the best I have.' Night before last he was called into the country. I found him at table in the morning, and with a heavy but hearty yawn said he, 'I had a drive last night, over a stony road, in a carriage without springs. I changed it, but was no better off, and I feel well pounded.' This was not a complaint, but an experience, and as soon as breakfast was over, eaten as it was with all sorts of interruptions, he was ready for his visit to the Duchess of ——, and everybody else. He eats little, and as if almost unconscious of the function. In this he constantly reminds me of ——.

"He receives a great deal of money, I have heard. But he seems wholly regardless of money, and, as I have further heard, it is only lately that he had begun to accumulate property. He is paid at the visit or consultation, which saves him from one of the most inconvenient offices, charging and collecting fees. We feel both the inconvenience and loss in America. I have seen fees paid him. It is when the patient is leaving him and by offering the hand for farewell the fee is deposited in his. I really think if he were subjected to our system he would get no money at all. 'At night,' said a patient of his, whom he sent to me when she came to America, 'his pockets are emptied. He knows nothing of their contents before, and so his money is cared for.' I said his meals are often interrupted. His butler brings in cards, notes, letters. 'There,' says he, and lays by note after note. Then two or three ladies come in. If he be not in, down they sit on the sofa, and take up books or newspapers. Then gentlemen, with or without ladies, appear. They are always asked to table by Miss ——, his sister-in-law, or somebody else. When the Professor is at table he places them. But he is reading and eating, or giving bread to a spotted Danish coach-dog named Billy, of fine size, and a universal pet. I feed him always. Professor S. talks to the comers. Then learns of strangers what they want, gets their residence if visits are wanted, or goes into a room hard by and sees them alone. His house is very large and full of rooms, and always seems inhabited. At length he gets ready to go out. 'Come away,' says he to me. I run up to put on a different coat, to get hat, &c., and always find him hat on at the door, ready to run down the steps for the morning's work. This is the way every day. He wears a narrow-brimmed hat, and puts it on well back, and shows his whole face and part ot his head. His dress is always black, with a remarkably nicely-arranged white neck-cloth, with a very carefully-made bow in front. So you see he is always dressed. I think, M., you would want to give the hat a different set. You could not improve the rest of the toilet. Now, is it not a great privilege to be the inmate of such an establishment as this? Is it not a thing to prize, to be the companion of a man so wholly devoted to others, and yet who is so cheerful, so constantly happy himself? You are admitted by such a man into the society of his thought and of his act. He always talks to the purpose, and yet he is the least of a formalist of any man with whom I have been acquainted. He has large information, for he is habitually an observer and a student; and yet he has no pedantry, no obtrusion of learning for its (or rather his) own sake, but that his companions may be helped by what he knows. He is almost daily making new observations, discovering something new, or using the known in a new way. And yet he is not in the remotest degree a dogmatist. It is not to support a doctrine that he talks, but to afford you an opportunity to speak more fully of it, to get knowledge from it, or to aid you by the knowledge he communicates. I have been chiefly a questioner in the society of Professor S., and I always have got good answers. If he has no answer, if he cannot explain the unexplained in my own mind, he turns himself round in his coach, for it is in driving I have the best of his society, and says, 'I don't know, I cannot explain that.' He will add, 'I have had the same difficulty you have, and cannot clear it up.' One advantage has arisen out of this intercourse with Professor S., which declares itself to me every day. I am conscious of a daily review of my own professional life, of thought, of reading, and of study. I speak constantly of books, of cases, of results of treatment. Professor S. has read all, and infinitely more than I have, and yet how small is his study. 'Here is my study,' said he the other night, as I was passing his sleeping-room, on my way to bed; 'come along.' In I went. The room was small. There was his bed, and in place of a night-stand, there was at the head of the bed a book stand or case, with two or three shelves about a foot and a-half wide, filled with books. The filling took but few. Taking hold of a movable gas-burner he brought it forward, so that he could easily read on his pillow. 'Here,' said he, 'is my study. Here I read at night.' I only said, 'What a privilege it is to be able to read in this way. I never could;' and then 'Good-night.' I heard his night-bell almost every night."

In the spring of 1850 he had a serious attack of illness resulting from an abscess in the arm-pit. At the close of the college session of that year he made a short Continental tour, which he enjoyed very much, seeing the principal universities in France, Germany, Holland, and Belgium, and everywhere receiving great kindness and attention. On his return, in spite of the overwhelming number of his engagements and professional duties, he found time to make some inquiries regarding leprosy and leper-houses, and continue the study of astronomy, in which latter study he was greatly assisted by the late Sir William Keith Murray, of Ochtertyre. His professional income had now increased so largely, that he frequently returned the fees to many of his poorer patients. Sums of £30, £25, £10, and £5 were, in 1850 and 1851, returned to those who had sent them. To a clergyman he once wrote: "Many kindest thanks for the fee which you have sent me. But you must be so good as allow me to return it, as (with many others of my professional brethren) I do not think it right to take any professional remuneration from clergymen or their families. Give me your prayers, and I will value them far more." With the accumulated increase in his income, he embarked in various speculations. In April, 1853, Mr. Robert Chambers inserted a paper in *Chambers's Journal* from his own pen, on the subject of "Oil-Anointing." This subject had been suggested to him by Simpson's researches. When on a professional visit to Galashiels, Roxburghshire, Dr. Macdougall remarked to him the healthy state and robust appearance of many of the operatives in the large woollen factories in the town. They seemed all specially and strikingly exempt from consumption and scrofulous diseases. This led him to recommend "oil-anointing" in all cases of weakness and threatened consumption. On 1st March, 1853, he received the high honour of being elected a Foreign Associate of the Academy of Medicine, Paris.

Amusement was not altogether forgotten. At a series of *tableaux vivants* held in his house at Queen Street, the fifth in order represented the "Babes in the Wood," and was greeted with roars of laughter and applause. Dr. Simpson and his colleague, Dr. L. P., were 'the Babes.' They entered sucking oranges, and dressed as children—short dresses, pinafores, frilled drawers, white socks, and children's house-shoes. After wandering about a while, they began to weep, then lay down and died, to the great delight of the juveniles."

Acupressure, a new blood-stopping process, discovered by him, and founded on the temporary metallic compression of arteries, was first described to the Royal Society of Edinburgh, 19th December, 1859. Its use speedily spread, and its application became very general and successful. The instruments he proposed to use were very sharp-pointed, slender needles or pins of passive or non-oxidisable iron, headed with wax or glass, and passed freely through the walls and flaps of wounds. These needles would cause little disturbance or irritation in the wounded part. Towards the end of 1861, he had begun, like Dr. Chalmers when writing his article on "Christianity," to question himself as to the faith that was in him. The result was a change of thought and feeling, and it became apparent that religion was a greater power in his life. At the end of this year, he entered a sick-room saying, "My *first* happy Christmas; my *only* happy one." The death of one of his children, "wee Jamie," deepened these impressions, and for a time he continued to address religious meetings. But he still continued to work with the old earnestness, and his archæological studies were as dear to him as ever. In 1866 the Queen conferred a baronetcy upon him. This was the signal for congratulations from all quarters, and the *Lancet*, in commenting upon it, said: "The conferring of this distinction must give, we think, universal satisfaction. Sir James has long been foremost in his department of practice, and his name is associated with the discovery of that invaluable boon to suffering humanity —chloroform. This alone would entitle him to the honour he has received. Sir James Y. Simpson is distinguished as an obstetric practitioner, as a physiologist, as an operator, and as a pathologist of great research and originality. His reputation is European, and the honour is fully deserved." The death of his eldest son, Dr. David James Simpson, shortly after receiving this honour, cast a gloom over No. 52 Queen Street. The death of a daughter, Jessie, at the early age of seventeen, was a new stroke to him. "Standing by her coffin," says his biographer, "he was able, as when he knelt by those of Jamie and of Davie, to say, 'Even so, Father; not my will, but Thine be done;' and to hear Christ's very voice, 'What I do thou knowest not now, but thou shalt know hereafter.'" But his work, the true antidote for sorrow, was entered into as heartily as before. In 1866 he received the degree of Doctor of Civil Law from the University of Oxford.

During the years 1866-67 he was busier than ever with his archæological studies. When any discovery of consequence was made in any part of the country, the result was pretty sure to be communicated to him. He published his own most important contribution to the science of this subject under the title of "Archaic Sculpturings of Cups, Circles, &c., upon Stones and Rocks in Scotland, England, and other countries." On the 25th August, 1868, he received a note from Provost William Chambers telling him that it was the intention of the Lord Provost, Magistrates, and Council to present him with the freedom of the city. The presentation took place about a year after this date. In the autumn of 1868 he gained the close acquaintance of Mr. Spurgeon. That this acquaintance was useful and valuable on both sides is seen from one of Mr. Spurgeon's notes, written after Sir James had made a professional call upon his wife. "I am writing far into the night to tell friends how my dear wife has sped. That dear angel of mercy, Sir James Simpson, has been very successful, as usual, and the operation is well over; patient, very patient, and in good spirits. If you know ten thousand eloquent men in Scotland, I would give them work for the next hundred years, viz., to praise the Lord for sending to us such a man, so skilful and so noble a doctor."

Early in 1870 it became clear to his familiar friends that he was more easily "knocked up than usual." In February Simpson had some severe attacks of illness, and by the end of the month he was quite prostrate. Those who were privileged to be present, said it was a wonderful sick-room. He would say: "I have not lived so near to Christ as I desired to do. I have had a busy life, but have not given so much time to eternal things as I should have sought. Yet I know it is not my merit I am to trust to for eternal life. Christ is all." Then he added with a sigh, "I have not got far on in the divine life." A friend said we are complete in Him. "Yes, that's it," he replied with a smile. "The hymn expresses my thought :

'Just as I am, without one plea,
But that Thy blood was shed for me.'

I *so* like that hymn." The days passed on. "The brother," says his biographer, "who had watched over him so tenderly in childhood, helped him in youth, and rejoiced in all the successes of his bright career, spent with him his last night but one on earth. He sat on the pillow with Sir James's head on his knee, on which he had been dandled in childhood, hearing ever and anon throughout that night's silent watching, the touching words, soft, and low, and slow, as if a weary sick child spoke, 'O Sandy, Sandy!'" On 6th May, 1870, he died without a struggle. A grave for the great departed in Westminster Abbey was offered to the family, but knowing Sir James's wish, this was declined, and the funeral took place to Warriston Cemetery on the 13th May. The funeral was one of the largest ever witnessed in Edinburgh. It was computed that over thirty thousand persons were present, either as spectators or as taking part in the procession.

GALLERY OF GREAT INVENTORS AND DISCOVERERS.

> "All the inventions that the world contains,
> Were not by reason first found out nor brains;
> But pass for theirs who had the luck to light
> Upon them by mistake or oversight."—BUTLER.

ALMOST all useful discoveries, it has been remarked, have been made not by the brilliancy of genius, but by the right direction of the mind to one object. In all trades, in all professions, success can be expected only from undivided attention. This common-sense view of things, a little different from that in the motto given above, is what we should adopt as we travel through the world of invention and discovery.

ROGER BACON.

Roger Bacon, a learned English monk of the Franciscan order, who flourished in the thirteenth century, was born near Ilchester, in Somersetshire, in 1214, and was descended of a very ancient and honourable family. He received the first tincture of letters at Oxford, where, having gone through grammar and logic, the dawnings of his genius gained him the favour and patronage of the greatest lovers of learning, and such as were equally distinguished by their high rank and the excellence of their knowledge. It is not very clear, says the "Biographia Britannica," whether he was of Merton College or of

Brazenose College, and perhaps he studied at neither, but spent his time at the public schools.

He went early over to Paris, where he made still greater progress in all parts of learning, and was looked upon as the glory of that university and an honour to his country. At Paris he did not confine his studies to any particular branch of literature, but endeavoured to comprehend the sciences in general fully and perfectly by a right method and constant application.

When he had attained the degree of Doctor, he returned again to his own country, and, as some say, took the habit of the Franciscans in 1240, when he was about twenty-six years of age; but others assert that he became a monk before he left France. After his return to Oxford, he was considered by the greatest men of that university as one of the ablest and most indefatigable inquirers after knowledge that the world had ever produced; and therefore they not only showed him all due respect, but likewise, conceiving the greatest hopes from his improvements in the method of study, they generously contributed to his expenses, so that he was enabled to lay out, within the compass of twenty years, no less than £2000 in collecting curious authors, making trials of various kinds, and in the construction of different instruments for the improvement of useful knowledge.

But if this assiduous application to his studies, and the stupendous progress he made in them, raised his credit with the better part of mankind, it excited the envy of some, and afforded plausible pretences for the malicious designs of others. It is very easy to conceive that the experiments he made in all parts of natural philosophy and the mathematics must have made a great noise in an ignorant age, when scarcely two or three men in a whole nation were tolerably acquainted with those studies, and when all the pretenders to knowledge affected to cover their own ignorance by throwing the most scandalous aspersions on those branches of science which they either wanted genius to understand, or which demanded greater application to acquire than they were willing to bestow. They gave out, therefore, that mathematical studies were in some measure allied to those magical arts which the Church had condemned, and thereby brought suspicion upon men of superior learning. It was owing to this suspicion that Bacon was restrained from reading lectures to the young students in the university, and at length closely confined and almost starved, the monks being afraid lest his writings should extend beyond the limits of his convent, and be seen by any besides themselves and the Pope. But there is great reason to believe that though his application to the occult sciences was their pretence, the true cause of his ill-usage was, the freedom with which he had treated the clergy in his writings, in which he spared neither their ignorance nor their want of morals.

Notwithstanding this harsh treatment, his reputation continued to spread over the whole Christian world, and even Pope Clement IV. wrote him a letter desiring that he would send him all his works. This was in 1266, when our author was in the flower of his age; and to gratify his Holiness, he collected together, greatly enlarged, and arranged in some order, the several pieces he had written before that time, and sent them the next year by his favourite disciple John of London, or rather of Paris, to the Pope. This collection, which is the same as he entitled "Opus Magnus," or his great work, is yet extant, and was published by Dr. Jebb in 1773.

It is said that this learned book procured Roger Bacon the

favour of Pope Clement IV., and also some encouragement in the prosecution of his studies; but this could not have lasted long, as that Pope died soon after, and then we find our author under fresh embarrassments from the same cause as before; but he became in more danger as the general of his order, having heard his cause, ordered him to be imprisoned. This is said to have happened in 1278; and to prevent his appealing to Pope Nicholas III., the general procured a confirmation of his sentence from Rome immediately, but it is not very easy to say on what pretences. It is certain that his sufferings for many years must have brought him low, since he was sixty-four years of age when he was first put in prison, and deprived of the opportunity of prosecuting his studies, at least in the way of experiment. That he was still indulged in the use of his books appears very clearly from the great use he made of them in the learned works he composed.

He was not released from prison till the latter end of the reign of Pope Nicholas IV., when he owed his freedom to the interposition of some noblemen. He returned to Oxford, where, at the request of his friends, he composed "A Compendium of Theology," which seems to have been his last work, and of which there is a copy in the royal library.

He spent the remainder of his days in peace, and died in the college of his order on the 11th of June, 1292, as some say, or in 1294, as others assert, and was interred in the Church of the Franciscans. The monks gave him the title of "Doctor Mirabilis," or the "Wonderful Doctor," which he deserved in whatever sense the phrase is taken.

He was certainly the most extraordinary man of his time. He was a perfect master of the Latin, Greek, and Hebrew, and has left posterity such indubitable marks of his critical skill in them, as might have secured him a very high character, if he had never distinguished himself in any other branch of literature. In all branches of the mathematics he was well versed. In mechanics particularly, the learned Dr. Friend says, that a greater genius had not arisen since the days of Archimedes. He comprehended likewise the whole science of optics with accuracy, and is very justly allowed to have understood both the theory and practice of those discoveries which have bestowed such high reputation on those of our own and other nations who have brought them into common use. In geography, also, he was admirably well skilled, as appears from a variety of passages in his works, which was the reason that induced the judicious Hakluyt to transcribe a large discourse out of his writings into his collection of travels.

But his skill in astronomy was even more remarkable, since it appears that he not only pointed out the error which occasioned the reformation in the calendar, and the distinction between the old style and the new, but also offered a much more effectual and perfect reformation than that which was made in the time of Pope Gregory XIII.

He was so thoroughly acquainted with chemistry, at a time when it was scarcely known in Europe, and principally cultivated among the Arabians, that Dr. Friend ascribes the honour of introducing it to him, who speaks in some part or other of his works of almost every operation now used in chemistry.

Three capital discoveries, or attempted discoveries, of his deserve to be particularly considered. The first is the invention of gunpowder, which, however confidently ascribed to others, was unquestionably known to him, both as to its ingredients and effects. The second is that which goes under the name alchemy, or the art of transmuting metals, of which he has left many treatises, some published and some still remaining in MS., which, whatever they may be thought of now, contain a multitude of curious and useful passages independently of their principal subject. The third discovery in chemistry, not so deserving of the reader's attention, was the tincture of gold for the prolongation of life, of which, Dr. Friend says, he has given hints in his writings, and has said enough to show that he was no pretender to this art, but understood as much of it as "any of his successors."

As to the vulgar imputation on his character of his leaning to magic, it was utterly unfounded, and the ridiculous story of his making a brazen head, which spoke and answered questions, is a calumny indirectly fathered upon him, having been originally imputed to Robert Grosseteste, Bishop of Lincoln. That he had too high an opinion of judicial astrology, and some other arts of that nature, was not so properly an error of his as of the age in which he lived; and considering how few errors, among the many which infected that age, appear in his writings, it may be easily forgiven.

WILLIAM LEE.

There is a singular confusion pervading the early history of the stocking-frame: persons, places, and dates are all jumbled up together in the accounts given of the inventor and the invention, and these accounts it is difficult to reconcile, unless we implicitly believe the evidence of a painting which long adorned the Stocking-weavers' Hall in Redcross Street, London. This portrait represented a man in collegiate costume, in the act of pointing to an iron stocking-frame, and addressing a woman who is knitting with needles by hand. The picture bore the following inscription: "In the year 1589, the ingenious William Lee, A.M. of St. John's College, Cambridge, devised this profitable art for stockings (but, being despised, went to France), yet of iron to himself, but to us and to others of gold; in memory of whom this is here painted."

In Deering's "Account of Nottingham" we learn that William Lee (whose name is sometimes written Lea) was a native of Woodborough, a village about seven miles from Nottingham. He was heir to a considerable freehold estate, and a graduate of St. John's College, Cambridge. It is said that he fell in love with a young country girl, who during his visits paid more diligent attention to her work, which was knitting, than to the fond speeches of her lover. He endeavoured, therefore, to invent a machine which might facilitate and forward the operation of knitting, and by this means furnish the object of his affections with more leisure to converse with him. Beckmann says: "Love indeed is fertile in inventions, and gave rise, it is said, to the art of painting; but a machine so complex in its parts, and so wonderful in its effects, would seem to require longer and greater reflection, more judgment, and more time and patience than could be expected in a lover. But even if the case should appear problematical, there can be no doubt in regard to the inventor, whom most of the English writers positively assert to have been William Lee." Deering expressly states that Lee made the first loom in the year 1589, the date inscribed on the picture.

But this is not the only version of the story. Another one states that Lee was expelled from the university for marrying contrary to the statutes. He had no fortune, and his wife was forced to contribute to their joint support by knitting. Lee, while watching the movement of her fingers, conceived the happy idea of imitating those movements by a machine. According to a third version, Lee, while yet unmarried, excited the contempt of his mistress by contriving a machine to imitate the primitive process of knitting, and was rejected by her. And a fourth account, slightly resembling the first tradition, exhibits Lee in a very unamiable light. It is said that he had taken a pique against a townswoman with whom he was in love, and who, it seems, disregarded his passion. She got her livelihood by knitting stockings, and with the unamiable view of depreciating her calling, he constructed the stocking-frame. He first worked at it himself, and afterwards taught his brothers and others of his relations.

All these accounts agree that the stocking-frame was invented by Lee. A writer in the *Quarterly Review*, 1816, however, cautiously observes: "This painting might give rise to the story of Lee's having invented the machine to facilitate the labour of knitting, in consequence of falling in love with a young country girl, who, during his visits, was more attentive to her knitting than to his proposals; or the story may, perhaps, have suggested the picture."

The story of Lee's after-life corroborates his being the inventor. He is mentioned as such in the petition of the stocking-weavers of London, to allow them to establish a guild. It is related that he practised his new invention some time at Calverton, near Nottingham. After remaining there for five years, he applied to Queen Elizabeth for countenance and support. She neglected him, and so did her successor, James I.; so Lee in disgust transferred himself and his machines to France, where Henri IV. and his gracious minister, Sully, gave the inventor a welcome reception. After the assassination of Henri, Lee shared in the persecution of the Protestants, and is said to have died in great distress. Some of his workmen made their escape to England, and under one Aston, who had been apprenticed to Lee, established the stocking manufacture permanently in England.

Lee's invention was important. It not only enabled our ancestors to discard their former inelegant hose, but it likewise caused the English manufactures to excel all of foreign production, and to be therefore eagerly sought after. Our makers soon exported vast quantities of silk stockings to Italy, and these so long maintained their superiority, that Keyslar, in his "Travels through Europe," as late as the year 1730, remarks: "At Naples, when a tradesman would highly recommend his silk stockings, he protests they are right English."

MARQUIS OF WORCESTER.

When this distinguished nobleman first published his "Century of Inventions," he was regarded by the public as at best a visionary projector, if not an absolute relator of falsehoods.

The Marquis, who had sacrificed his fortune in scientific pursuits, wished to obtain the encouragement of the King or of the Parliament, and offered to carry his grand projects into effect *gratis*. In a dedication to the King, speaking of the list of his inventions, he says: "If it might serve to give aim to your Majesty how to make use of my poor endeavours, it would crown my thoughts, who am neither covetous nor ambitious, but of deserving your Majesty's favour, upon my own cost and charges; yet according to the old English proverb, 'It is a poor dog not worth whistling after.' Let but your Majesty approve, and I will effectually perform to the height of my undertaking; vouchsafe but to command, and with my life and fortune I shall cheerfully obey, and *maugre* envy, ignorance, and malice, ever appear your Majesty's passionately devoted, or otherwise disinterested, subject and servant,

"WORCESTER."

In a second dedication to the members of the two Houses of Parliament, he states that he had already spent more than £10,000 in maturing his discoveries for the public good. He speaks of them with that modest confidence so inseparable from transcendent talents :

"The treasures buried under these heads," he says, "both for war, peace, and pleasure, being inexhaustible, I beseech you pardon me if I say so; it seems a vanity, but comprehends a truth; since no good spring but becomes the more plentiful, by how much more it is drawn; and the spinner to weave his web is never stinted, but enforced.

"The more, then, that you shall be pleased to make use of my inventions, the more inventive shall you ever find me, one invention begetting still another, I more and more improving my ability to serve my King and you; and as to my heartiness therein, there needs no addition, nor to my readiness a spur. And therefore, my lords and gentlemen, be pleased to begin, and desist not from commanding me, till I flag in my obedience and endeavours to serve my King and country.

> 'For certainly you'll find me breathless first t' expire,
> Before my hands grow weary, or my legs do tire.'"

It may be observed, that however much his work was slighted in his own day, it is pretty clear that the Marquis suggested the first idea of the *steam-engine*, and that in like manner he evidently hints at the *telegraph*, the *torpedo*, and at the *velocipede*. And it is not improbable that in his 15th "Scantling," "A boat driving against wind and tide," he had an eye to steam navigation.

In his "Century of Inventions," the manuscript of which, by the way, dates from 1655, he describes a steam apparatus by which he raised a column of water to the height of 40 feet. This, under the name of "Fire-waterwork," appears actually to have been at work at Vauxhall in 1656.

PRINCE RUPERT.

Prince Rupert, third son of the King of Bohemia, by the Princess Elizabeth, eldest daughter of James I. of England, was born in 1619, and educated, like most German princes, for the army; and those who have been least inclined to favour him admit that he was well adapted, both by natural abilities and acquired endowments, to form a great commander. On the commencement of the rebellion, which happened when he was scarcely of age, he offered his services to King Charles, and throughout the whole war behaved with great intrepidity.

On the Restoration he was invited to return to England, and had several offices conferred upon him. After the display of considerable ability in the Dutch wars of the reign of Charles II., he withdrew into retirement, mostly at Windsor Castle, of which he was governor, and spent a great part of his time in the prosecution of chemical and philosophical experiments, as well as the practice of mechanical arts. He delighted in making locks for fire-arms, and was the inventor of a composition called from him Prince's metal.

He communicated to the Royal Society his improvements upon gunpowder, by refining the several ingredients, and making it more carefully, which augmented its force, in comparison of ordinary powder, in the proportion of ten to one. He also acquainted them with an engine he had contrived for raising water, and sent them an instrument for casting any platform into perspective, and for which they deputed a select committee of their members to return him their thanks.

He was the inventor of a gun for discharging several bullets with the utmost speed, facility, and safety; and the Royal Society received from His Highness the intimation of a certain method of blowing up rocks in mines and other subterraneous places. Dr. Hooke has preserved another invention of his for making hail shot of all sizes. He devised a particular kind of screw, by means of which observations taken by a quadrant at sea were secured from receiving any alteration by the unsteadiness of the observer's hand or through the motion of the ship. It is said that he had also, among other secrets, that of melting or running black lead like a metal into a mould, and reducing it again into its original form.

But there is one invention of which he has the credit, which requires more particular notice. Besides being mentioned by foreign authors, with applause for his skill in painting, he was considered as the inventor of mezzotinto, owing, as it is said, to the following casual occurrence. Going out early one morning during his retirement at Brussels, he observed the sentinal at some distance from his post very busy doing something to his piece.

The Prince asked the soldier what he was about?

He replied that the dew had fallen in the night and made his fusil rusty, and that he was scraping and cleaning it.

The Prince, looking at it, was struck with something like a figure eaten into the barrel, with innumerable little holes closed together like friezed work on gold or silver, part of which the fellow had scraped away. The Prince immediately conceived that some contrivance might be found to cover a brass plate with such a grained ground of fine pressed holes, which would undoubtedly give an impression all black; and that by scraping away proper parts, the smooth superficies would leave the rest of the paper white. Communicating his idea to Walleraut Vaillant, a reputable painter then in the neighbourhood of Brussels, they made several experiments, and at last invented a steel roller with projecting points or teeth like a file, which effectually produced the black ground, and which being scraped away or diminished at pleasure left the gradations of light.

Such was the invention of mezzotinto, according to Lord Orford, Mr. Evelyn, and Mr. Vertue, though in all fairness we must state that it has been disputed by some, and the credit assigned to a German, Von Tregen, whose early works bear the date of 1642.

The earliest of Rupert's engravings in mezzotinto that is now extant is dated 1658. It is a half-length figure from Spagnoletto: the subject, an executioner holding a sword in one hand, and in the other a head, which is probably intended for that of John the Baptist.

Prince Rupert died at his house in Spring Gardens, 29th November, 1682, and was interred in Henry VIII.'s chapel, regretted as one whose aim in all his actions and all his accomplishments was the public good.

SIR SAMUEL MORLAND.

As a machinist, Sir Samuel Morland, who was born about 1625, deserves more respect than has hitherto been paid to him. He invented the speaking-trumpet, the fire-engine, a capstan to heave up anchors, and two arithmetical machines, of which he published a description under the title of "The Description and Use of Two Arithmetical Instruments; together with a Short Treatise, explaining the Ordinary Operations of Arithmetic, &c., presented to His Most Excellent Majesty Charles II. by S. Morland in 1692." This work, which is exceedingly rare, but of which there is a copy in the Bodleian, which bears date 1673, 8vo, is illustrated with twelve plates, in which the different parts of the machine are exhibited; and whence it appears that the four fundamental rules in arithmetic are very readily worked, and to use the author's own words, "without charging the memory, disturbing the mind, or exposing the operations to any uncertainty." That these machines were at the time brought into practice there seems no reason to doubt, as by an advertisement prefixed to the work it appears that they were manufactured for sale by Humphrey Adamson, who lived with Jonas Moore, Esq., in the Tower of London.

The speaking-trumpet is said to have been invented by Sir Samuel Morland in 1670. His trumpet was of the same form as that now in use—that is to say, it was a truncated cone, with an outward curve or lip at the opening.

"The theory of the action of this instrument," says one writer, "has never been thoroughly explained; but it is supposed that the sides of the tube throw the sound back and back in various reflections, until ultimately the waves quit the instrument in parallel lines. It does not seem to depend on the vibration of the instrument."

JOHN FLAMSTEAD.

John Flamstead, the great astronomer, who was a contemporary of Sir Isaac Newton, was led to the study of astronomy by perusing Sacrobosco's work, "De Sphæra." He prosecuted his studies with so much assiduity as to be appointed first Astronomer-Royal. His great work is entitled "Historia

Cœlestis Britannicæ." This publication contains his famous catalogue of the fixed stars, the first trustworthy one ever made; the immense mass of celestial observations of which the catalogue was the fruit, or rather the first fruit; and a full account of his methods of observation and his instruments.

Like most men of superior learning in his day, Flamstead had the reputation among the lower orders of being able to foretell events. In this persuasion, a poor washerwoman of Greenwich, who had been robbed at night of a parcel of linen, came to him, and with great anxiety requested him to use his art, to let her know where the linen was, and who had robbed her. The doctor, who was a humourist, thought he would indulge himself in the joke; he bade the poor woman stay, and he would see what he could; perhaps he might let her know where she might find it; but who the persons were he would not undertake to inform her, for as she could have no positive proof to convict them, it would be useless. He then set about drawing circles, squares, &c., to amuse her; and after some time, told her if she would go into a particular field, she would find the whole bundled up in a part of the ditch. The woman repaired there immediately, and found it. She came back with great haste and joy to thank the doctor, and offered him half-a-crown as a token of her gratitude, that being as much as she could afford. The doctor, more surprised than the woman, told her, "Good woman, I am heartily glad you have found your linen; but I assure you I knew nothing of it, and intended only to joke with you, and then to have read you a lecture on the folly of applying to any person to know events not in the human power to tell. But I see Satan has a mind I should deal with him: I am determined, however, I will not; so never come nor send any one to me any more, on such occasions, for I never will attempt such an affair again whilst I live."

JOHN HARRISON.

John Harrison, the inventor of the time-keeper which procured him the reward of the Board of Longitude, was the son of a carpenter in Yorkshire, and assisted his father in the business until he was twenty years of age. Occasionally, however, he was employed in measuring land, and mending clocks and watches. He was from his childhood attached to any wheel machinery; and when he lay ill in his sixth year, he had a watch placed open upon his pillow, that he might amuse himself by contemplating the movement. Though his opportunities of acquiring knowledge were very few, he eagerly improved every incident for information. He frequently passed whole nights in drawing or writing; and he always acknowledged his obligations to a neighbouring clergyman for lending him a manuscript copy of Professor Sanderson's Lectures which he carefully and neatly transcribed, with all the diagrams.

On the reward being offered, in the 14th of Queen Anne, for discovering the longitude, Harrison's attention was drawn to the subject, and he began to consider how he could alter a clock, which he had previously made, so that it might not be subject to any irregularities, occasioned by the difference of climates and the motions of a ship. These difficulties he surmounted; and his clock having answered his expectations in a trial, attended with very bad weather upon the river Humber, he was advised to carry it to London, in order to apply for the parliamentary reward. He first showed it to several members of the Royal Society, who gave him a certificate that his machine for measuring time promised a very great and sufficient degree of exactness. In consequence of this certificate, the machine, at the recommendation of Sir Charles Wager, was put on board a man-of-war in 1736, and carried, with Mr. Harrison, to Lisbon and back again, when its accuracy was such, that the Commissioners of the Board of Longitude gave him £500, and recommended him to proceed. He made two others afterwards, each of which was an improvement on the preceding, and he now thought he had reached the *ne plus ultra* of his attempts; but in an endeavour to improve pocket watches, he found the principles he applied to surpass his expectations so much as to encourage him to make his fourth time-keeper, which was in the form of a pocket-watch, about sixteen inches in diameter, and was finished in 1759. With this time-keeper his son made two voyages, the one to Jamaica, and the other to Barbadoes, in both which experiments it corrected the longitude within the nearest limits required by the Act of Parliament; and the inventor, at different times, though not without considerable trouble, received the promised reward of £20,000.

GEORGE GRAHAM.

George Graham, clock and watch maker, the most ingenious artist of his time, was born at Horsgills, in the parish of Kirklinton in Cumberland, in the year 1675.

In 1688 he came up to London, and was put apprentice to a person in that profession; but after being some time with his master, he was received, purely on account of his merit, into the family of the celebrated Mr. Tompion, who treated him with a kind of paternal affection as long as he lived.

That Mr. Graham was, without competition, the most eminent of his profession, is but a small part of his character. He was the best general mechanic of his time, and had a complete knowledge of practical astronomy; so that he not only gave to various movements for measuring time a degree of perfection which had never before been attained, but invented several astronomical instruments, by which considerable advances have been made in that science; he also effected great improvements in those which had before been in use; and, by a wonderful manual dexterity, constructed them with greater precision and accuracy than any other person in the world.

A great mural arch in the Observatory at Greenwich was made by Dr. Halley, under Mr. Graham's immediate inspection, and divided by his own hand; and from this incomparable original, the best foreign instruments of the kind are copies made by English artists. The sector by which Dr. Bradley first discovered two new motions in the fixed stars was of his invention and fabrication. He comprised the whole planetary system within the compass of a small cabinet, from which, as a model, all the modern orreries have been constructed. And when the French Academicians were sent to the north to make observations for ascertaining the figure of the earth, Mr. Graham was thought the fittest person in Europe to supply them with instruments; by which means they finished their operations in one year, while those who went to the south, not being so well furnished, were very much embarrassed and retarded in their operations.

Mr. Graham was many years a member of the Royal Society, to which he contributed several ingenious and important discoveries, chiefly on astronomical and philosophical subjects; particularly a kind of horary alteration of the magnetic needle, a quicksilver pendulum, and many curious particulars relating to the true length of the simple pendulum, upon which he continued to make experiments till almost the year of his death, which happened on the 20th of November, 1751, in his house in Fleet Street.

His temper was not less communicative than his genius was penetrating; and his principal view was the advancement of science and the benefit of mankind. As he was perfectly sincere, he was above suspicion; as he was above envy, he was candid; and as he had a relish for true pleasure, he was generous. He frequently lent money, but could never be prevailed upon to take any interest; and for that reason he never placed out any money upon Government securities. He had bank-notes which were thirty years old in his possession when he died; and his whole property, except his stock-in-trade, was found in a strong-box, which, though less than would have been heaped together by avarice, was yet more than would have remained to prodigality.

JAMES FERGUSON.

James Ferguson, the eminent practical philosopher and astronomer, was born in a humble station at Keith, a small village in Scotland, in the year 1710. He learned to read by merely listening to the instructions which his father communicated to an elder brother. He was afterwards sent for about three months to the grammar school at Keith; and this was all the scholastic education he ever received. His taste for mechanics appeared when he was only about seven or eight years of age. By means of a turning-lathe and a knife, he constructed machines that served to illustrate the properties of the lever, the wheel and axle. Of these machines, and the mode of their application, he made rough drawings with a pen, and wrote a brief description of them. Unable to subsist without some employment, he was placed with a neighbouring farmer, and was occupied for some years in the care of his sheep. In this situation he commenced the study of astronomy, devoting a great part of the night to the contemplation of the heavens; while he amused himself in the day-time with making models of spinning-wheels, and other machines which he had an opportunity of observing. By another farmer, in whose service he was afterwards engaged, he was much encouraged in his astronomical studies, and enabled, by the assistance that was afforded him in his necessary labour, to reserve part of the day for making fair copies of the observations which he roughly sketched out at night. In making these observations, he lay down on his back, with a blanket about him, and by means of a thread strung with small beads, and stretched at arm's length between his eye and the stars, he marked their positions and distances. The master who thus kindly favoured his search after knowledge, recommended him to some neighbouring gentlemen, one of whom took him into his house, where he was instructed by the butler in decimal arithmetic, algebra, and the elements of geometry! Being afterwards deprived of the assistance of this preceptor, he returned to his father's house, and, availing himself of the information derived from Gordon's *Geographical Grammar*, constructed a globe of wood, covered it with paper, and delineated upon it a map of the world; he also added the meridian ring and horizon, which he graduated; and by means of this instrument, which was the first he had ever seen, he came to solve all the problems in Gordon. His father's contracted circumstances obliged him again to seek employment; but the service into which he entered was so laborious as to affect his health. For his amusement in this enfeebled state, he made a wooden clock, and also a watch, after having once seen the inside of such a piece of mechanism. His ingenuity obtained for him new friends, and employment suited to his taste, which was that of cleaning clocks, and drawing patterns for ladies needlework; and he was thus enabled not only to supply his own wants, but to assist his father. Having improved in the art of drawing, he was induced to draw portraits from the life with Indian ink on vellum. This art, which he practised with facility, afforded him a considerable subsistence for several years, and allowed him leisure for pursuing those favourite studies which ultimately raised him to eminence.

"My taste for mechanics," says Mr. Ferguson, in a sketch of his own life, "arose from an odd accident. When about seven or eight years of age, a part of the roof of the house being decayed, my father applied a prop and lever to an upright spar to raise it to its former situation; and, to my great astonishment, I saw him, without considering the reason, lift up the ponderous roof as if it had been a small weight. I attributed this, at first, to a degree of strength that excited my terror as well as wonder; but thinking further of the matter, I recollected that he had applied his strength to that end of the lever which was farthest from the prop, and finding, on inquiry, that this was the means by which the seeming wonder was effected, I began making levers (which I then called bars), and by applying weights to them different ways, I found the power given by my bar was just in proportion to the lengths of the different parts of the bar on either side of the prop. I then thought it was a great pity that by means of this bar a weight could be raised but a very little way. On this I soon imagined that, by pulling round a wheel, the weight might be raised to any height by tying a rope to the weight, and winding the rope round the axle of the wheel, and that the power gained must be just as great as the wheel was broader than the axle was thick, and found it to be exactly so, by hanging one weight to a rope put round the wheel, and another to the rope that coiled round the axle, so that in these two machines it appeared very plain that their advantage was as great as the space gone through by the working power exceeded the space gone through by weight; and this property, I thought, must take place in a wedge for cleaving wood, but then I happened not to think of the screw. I then wrote a short account of the machines, and sketched out figures of them with a pen, imagining it to be the first treatise of the kind that ever was written." So early did this philosopher's genius for mechanics first appear; and from such small beginnings did that knowledge spring for which he was afterwards so justly distinguished.

MATTHEW BOULTON.

Matthew Boulton, the partner of James Watt, also deserves mention here. He was born on the 3rd of September, 1728, at Birmingham, where his father carried on business as a hardwareman.

He received an ordinary education at a school in Deritend, and also acquired a knowledge of drawing and mathematics. At the age of seventeen, he effected some improvements in shoe-buckles, buttons, and several other articles of Birmingham manufacture.

The death of his father left him in possession of considerable property; and in order to extend his commercial operations, he purchased, about 1762, a lease of Soho, near Handsworth, about two miles from Birmingham, but in the county of Stafford,

It would scarcely be possible, says one writer, to select a more striking instance of the beneficial changes effected by the combined operations of industry, ingenuity, and commerce, than that which was presented by Soho after it had been some time in Mr. Boulton's possession. Previously it had been a bleak and barren heath, but it was soon diversified by pleasure grounds, in the midst of which stood Mr. Boulton's spacious mansion, and a range of extensive and commodious workshops capable of receiving over a thousand artisans.

To Mr. Boulton's active mind this country is eminently indebted for the manner in which he extended its resources, and brought into repute its manufacturing ingenuity. Water was an inadequate moving power in seconding his designs, and he had recourse to steam. The old engine on Savary's plan was not adapted for some purposes in which it was necessary that great power should be combined with delicacy and precision of action.

In 1769, Mr. Boulton having entered into communication with James Watt, who had obtained a patent for improvements in the steam-engine, Watt was induced to settle at Soho. In 1775, Parliament granted him a further extension of his patent for improvements in the steam-engine; and on his entering into partnership with Mr. Boulton, the Soho works soon became famous for their excellent engines. Not only was the steam-engine itself brought to greater perfection, but its powers were applied to a variety of new purposes. In none of these, perhaps, was the success so remarkable as in the machinery for coining, which was put in motion by steam. The coining apparatus was first set agoing in 1783, but it soon underwent important improvements, until it was at length brought to an astonishing degree of perfection. One engine put in motion eight machines, each of which stamped on both sides and milled at the edges from seventy to eighty-four pieces in a minute; and the eight machines together completed, in a style far superior to anything which had previously been accomplished, from 30,000 to 40,000 coins in an hour.

The manufacture of plated wares, of works in bronze and ormolu, such as vases, candelabra, and other ornamental articles, was successively introduced at Soho, and the taste and excellence which these productions displayed soon obtained for them an unrivalled reputation in every part of the world. Artists and men of taste were warmly encouraged, and their talents called forth, by Mr. Boulton's liberal spirit. The united labours of the two partners contributed to give that impulse to British industry which has never since ceased.

Mr. Boulton has been described by Playfair as possessing a most generous and ardent mind, to which was added an enterprising spirit that led him to grapple with great and difficult undertakings. "He was a man of address," continues the same writer, "delighting in society, active, and mixing with people of all ranks with great freedom and without ceremony." Watt, who survived Mr. Boulton, spoke of his deceased partner in the highest terms. He said, "To his friendly encouragement, to his partiality for scientific improvement, and to his ready application of them to the purposes of art, to his intimate knowledge of business and manufactures, and to his extended views and liberal spirit, may in a great measure be ascribed whatever success may have attended my exertions."

Mr. Boulton expended about £47,000 in the course of experiments on the steam-engine, before Watt perfected the construction and occasioned any return of profit.

JOSEPH BLACK.

Joseph Black, the famous chemist, was born in France in 1728. He was educated at the Universities of Glasgow and Edinburgh. In 1756 he was appointed Professor of Anatomy and Lecturer on Chemistry at Glasgow. It was during his residence at Glasgow that he made and established his discovery of latent heat.

The following most interesting account of one of the principal discoveries in modern science is from a biographical memoir prefixed by Professor Robinson to Dr. Black's Lectures :—

"It seems to have been between the years 1759 and 1763 that his speculations concerning *heat*, which had long occupied his thoughts, were brought to maturity. And when it is considered by what simple experiments, by what familiar observations, Dr. Black illustrated the laws of fluidity and evaporation, it appears wonderful that they had not long before been observed and demonstrated. They are, however, less obvious than might at first sight be imagined; and to have a clear and distinct conception of those seemingly simple processes of nature required consideration and reflection.

"If a piece of wood, a piece of lead, and a piece of ice are placed in a temperature much inferior to that of the body, and if we touch the piece of wood with the hand, it feels cold; if we touch the piece of lead, it feels colder still; but the piece of ice feels colder than either. Now, the first suggestion of sense is that we receive cold from the wood, that we receive more from the lead, and most of all from the ice; and that the ice continues to be a source of cold till the whole be melted. But an inference precisely the contrary to all this is made by him whose attention and reflection has been occupied with this subject. He infers that the wood takes a little heat from the hand, but is soon heated so much as to take no more. The lead takes more heat before it be as much satiated; and the ice continues to feel equally cold, and to carry off heat as fast as in the first moment till the whole be melted. This, then, was the inference made by Dr. Black.

"Boerhaave has recorded an interesting observation by Fahrenheit, namely, that water would sometimes grow considerably colder than melting snow without freezing, and would freeze in a moment when shaken or disturbed, and in the act of freezing give out many degrees of heat. Founded on this observation, it appears that Dr. Black entertained some vague notion or conjecture that the heat which was received by the ice during its conversion into water was not lost, but was still contained in the water. And he hoped to verify this conjecture, by making a comparison of the time required to raise a pound of water one degree in its temperature, with the time required to melt a pound of ice, both being supposed to receive the heat equally fast. And that he might ascertain how much heat was extricated during congelation, he thought of comparing the time required to depress the temperature of a pound of water one degree, with the time required for freezing it entirely. The plan of this series of experiments occurred to him during the summer season; but for want of ice, which he could not then procure, he had no opportunity of putting it to the test. He therefore waited impatiently for the winter.

"The winter arrived, and the decisive experiment was performed in the month of December, 1761. From this experiment, it appeared that as much heat was taken up by the ice during its liquefaction as would have raised the water 140 degrees in its temperature; and, on the other hand, that exactly the same quantity of heat was given out during the congelation of the water. But this experiment, the result of which Dr. Black eagerly longed for, only informed him how much heat was absorbed by the ice during liquefaction, was retained by the water while it remained fluid, and was again emitted by it in the process of freezing. But his mind was deeply impressed with the truth of the doctrine, by reflecting on the observations that presented themselves when a frost or thaw happened to prevail. The hills are not at once cleared of snow during the sunshine of the brightest winter day, nor are the pools suddenly covered with ice during a single frosty night. Much heat is absorbed and fixed in the water during the melting of the snow, and, on the other hand, while the water is changed into ice, much heat is extricated. During a thaw, the thermometer sinks when it is removed from the air and placed in the melting snow; during severe frost, it rises when plunged into freezing water. In the first case the snow receives heat, and in the last the water allows the heat to escape again. These were fair and unquestionable inferences, and now they appear obvious and easy. But although many ingenious and acute philosophers had been engaged in the same investigations, and had employed the same facts in their disquisitions, those obvious inferences were entirely overlooked. It was reserved for Dr. Black to remove the veil which hid this mystery of nature, and by this important discovery to establish an era in the progress of chemical science—one of the brightest, perhaps, which has yet occurred in its history."

The theory of *latent heat*—as Dr. Black called it was explained by him to the members of a literary society on the 23rd of April, 1762, and soon afterwards he laid before his students a detailed view of the extensive and beneficial effects of this habitude in the grand economy of nature. "From observing the analogy between the cessation of expansion by the thermometer during the liquefaction of the ice, and during the conversion of water into steam, Dr. Black having explained the one, thought that the phenomena of boiling and evaporation would admit of a similar explanation. He was so convinced of the truth of this theory, that he taught it in his lectures in 1761, before he had made a single experiment on the subject. At this period, his prelections on the subject of evaporation were of great advantage to James Watt, afterwards so distinguished, who made use of them in his application of steam power. Black's discovery, indeed, may be said to have laid the foundation of that great practical use of steam which has conferred so great a blessing upon the present age."

JOSEPH PRIESTLEY.

Joseph Priestley, a Dissenting divine, but more justly eminent as a philosopher, was born on the 18th of March, 1733, near Leeds.

At Warrington, where he occupied the post of tutor in an academy, he first began to acquire reputation as a writer in various branches of literature. In 1767, he gave to the world his *History of Electricity*. It is rather carelessly and hastily executed, but must have been of advantage to the science. Almost the whole of his historical facts are taken from the *Philosophical Transactions;* but at the end he gives a number of original experiments of his own. The most important of all his electrical discoveries was that charcoal is a conductor of electricity, and so good a conductor that it vies even with the metals themselves. This publication went through several editions, was translated into foreign languages, and procured him the honour of being elected a Fellow of the Royal Society, as one of his biographers says; but his election took place the year before, and about the same time the University of Edinburgh conferred on him the degree of Doctor of Laws.

His attention was now turned to the properties of fixed air He commenced his experiments on this subject in 1768, and the first of his publications appeared in 1772, in which he announced a method of impregnating water with fixed air. In the paper read to the Royal Society in 1772, which obtained the Copley medal, he gave an account of his discoveries; and at the same time announced the discovery of nitrous gas, and its application as a test of the purity and fitness for respiration of gases generally.

About this time also he showed the use of the burning lens in pneumatic experiments; he related the discovery and properties of muriatic acid gas; added much to what was known of the gas generated by putrefactive processes, and by vegetable fermentation; and he determined many facts relative to the diminution and deterioration of air, by the combustion of charcoal and the calcination of metal.

In 1774 he made a full discovery of dephlogisticated gas, which he produced from the oxides of silver and lead. This hitherto secret source of animal life and animal heat, of which Mayon had a faint glimpse, was unquestionably first exhibited by Dr. Priestley, though it was discovered about the same time by Scheele of Sweden.

In 1776 his observations on respiration were read before the Royal Society, in which he discovered that the common air inspired was diminished in quantity and deteriorated in quality by the action of the blood upon it, through the blood-vessels of the lungs; and that the florid red colour of arterial blood was communicated by the contact of air through the containing vessels.

But it is impossible to enumerate here all Dr. Priestley's discoveries; they are too numerous. "How many invisible fluids," says one writer, "whose existence evaded the sagacity of foregoing ages, has he made known to us! The very air we breathe he has taught us to analyse and examine, and to improve a substance so little known that even the precise effect of respiration was an enigma until he explained it. He first made known to us the proper food of vegetables, and in what the difference between these and animal substances consists. To him pharmacy is indebted for the method of making artificial mineral waters, as well as for a shorter method of preparing other medicines; metallurgy, for more powerful and cheap solvents; and chemistry for such a variety of discoveries as it would be tedious to recite—discoveries which new-modelled that science, and drew to it and to this country the attention of Europe."

Upon the life of Dr. Priestley, apart from science, we shall not touch. He died in America, on the 6th of February, 1804.

When the Council of the Royal Society honoured Dr. Priestley by the presentation to him of Sir Godfrey Copley's medal, on the 30th of November, 1733, Sir John Pringle, who was then president, delivered on the occasion an elaborate discourse on the different kinds of air, in which, after expatiating upon the discoveries of his predecessors, he pointed out the particular merits of Priestley's investigations. In allusion to the purification of a tainted atmosphere by the growth of plants, the president thus eloquently and piously expressed himself:—

" From these discoveries we are assured that no vegetable

grows in vain; but that, from the oak of the forest to the grass of the field, every individual plant is serviceable to mankind; if not always distinguished by some private virtue, yet making a part of the whole which cleans and purifies our atmosphere. In this the fragrant rose and deadly nightshade co-operate; nor is the herbage nor the woods that flourish in the most remote and unpeopled regions unprofitable to us, nor we to them, considering how constantly the winds convey to them our vitiated air, for our relief and for their nourishment. And if ever these salutary gales rise to storms and hurricanes, let us still trace and revere the ways of a beneficent Being, who, not fortuitously but with design, not in wrath but in mercy, shakes the water and the air together, to bury in the deep those putrid and pestilential effluvia which the vegetables on the face of the earth had been insufficient to consume."

In perusing the works of Priestley, it is impossible to fail being struck with his intense love of truth. In his scientific note-books, he registered every fact as it appeared to his senses; in his political and theological writings, he fearlessly states his opinions as they are brought out by his cross-examination of his own thoughts and meditations; and that liberty of independent thought which he claims for himself he determinedly demands for others. In his scientific career, his object was uniformly to question Nature by every possible experimental investigation, and to state its results as he obtained them.

"Whether we view him," says Professor Thomson, "as a pneumatic chemist, a theologian, an electrician, a historian, a politician, his writings bear the impress of an original mind, uncontrolled by any tendency to follow in beaten tracts, but constantly panting for new fields of investigation. It will ever remain a stain upon the name of England, that this noble-minded man, this honour to humanity, should have been compelled, by persecution on account of his religion and politics, to flee his native country."

JAMES HARGREAVES.

About a century has elapsed, says a popular writer, since a native of Lancashire, of very humble origin, began to devote his attention to the application of machinery to the preparation and spinning of raw cotton for weft. In the year 1760, or soon after, a *carding-engine*, not very different from that now in use, was contrived by James Hargreaves, an untaught weaver, living near Church, in Lancashire; and in 1767, the *spinning-jenny* was invented by the same person. This machine, as at first formed, contained eight spindles, which were made to revolve by means of bands from a horizontal wheel. Subsequent improvements increased the power of the *spinning-jenny* to eighty spindles, when the saving of labour which it thus occasioned, produced considerable alarm among those persons who had employed the old mode of spinning, and a party of them broke into Hargreaves' house and destroyed his machine. The great advantage of the invention was so apparent, however, that it was soon again brought into use, and nearly superseded the employment of the old spinning-machine. A second rising then took place of the persons whose labour was thus superseded by it. They went through the country destroying, wherever they could find them, both carding and spinning machines, by which means the manufacture was for a time driven away from Lancashire to Nottingham.

According to Hargreaves' own account, he derived the idea of the jenny from the following incident:—Seeing a hand-wheel with a single spindle overturned, he remarked that the spindle, which was before horizontal, was then vertical; and as it continued to revolve, he drew the roving of wool towards him into a thread. It then seemed to Hargreaves plausible that, if something could be applied to hold the rovings as the finger and thumb did, and that contrivance to travel backwards on wheels, six, or eight, or even twelve threads, from so many spindles, might be spun at once. This was done, and succeeded; but Hargreaves, driven by mobs, as we have described, to Nottingham, unable to bear up against such ill-treatment, there died in obscurity and distress. Before his death, he gave the property of his jenny to the Strutts, who thereupon laid the foundation of their industrial success and opulence.

JOSIAH WEDGWOOD.

Few men have laboured with such success to elevate and refine their art as Josiah Wedgwood, "the Father of the Potteries," as he has been called, and the first of a long line of Staffordshire potters who have applied the highest science and the purest art to the improvement of their commercial enterprise.

Wedgwood was born on the 12th of July, 1730, at Burslem in Staffordshire, and was the younger son of a potter. He received a very limited education, for scarcely any person in Burslem learned more than mere reading and writing, until about 1750, when some benevolent persons endowed the free school for instructing youth to read the Bible, write a fair hand, and know the primary rules of arithmetic. On his father's death, his property, consisting chiefly of a small entailed estate, descended to the eldest son, and Josiah was left at an early period of life to lay the foundation of his own fortune. This he did most substantially by applying his attention to the pottery business, which it is not too much to say he brought to the highest perfection, and established a manufacture that has opened a new source of extensive commerce before unknown to this or any other country.

His many discoveries of new species of earthenwares and porcelains, his studied forms and chaste style of decoration, and the correctness and judgment with which all his works were executed under his own eye, and by artists for the most part of his own forming, turned the current in this branch of commerce. Before his time, England imported the finer earthenwares; but since Wedgwood's day, she has exported them to an enormous extent.

The first ware by which Wedgwood attained to great celebrity was the improved cream-coloured ware. Of this new article he presented some specimens to Queen Charlotte, who immediately ordered a complete table service; and was so pleased with its execution that she appointed Wedgwood her potter, and commanded that the ware should be called "Queen's Ware." It has a dense and durable substance, covered with a brilliant glaze, and is capable of bearing uninjured sudden alterations of heat and cold. "It was from the first," says the late Mr. Timbs, "sold at a cheap rate, and the addition of embellishments very little enhanced the cost; first, a coloured edge or painted border was added to the queen's ware; and lastly, printed patterns, which covered the whole surface. Nor was this beautiful ware confined to England; for M. Faujas de Saint Ford shows how widely the fame of Wedgwood's pottery had spread before 1792, when "in travelling from Paris to Petersburg, from Amsterdam to the farthest part of Sweden, and from Dunkirk to the extremity of the south of France, one is served at every inn upon English ware. Spain, Portugal, and Italy are supplied with it; and vessels are loaded with it for the East Indies, the West Indies, and the continent of America." England is mainly indebted to Wedgwood for the extraordinary improvement and rapid extension of this branch of industry. Before his time, our potteries produced only inferior fabrics, easily broken or injured, and totally devoid of taste as to form or ornament.

When the Portland or Barberini vase was offered for sale, Wedgwood, with the view of copying it, endeavoured to purchase it, and for some time continued to offer an advance upon each bidding of the Duchess of Portland; until at length, his motives being ascertained, he was offered the loan of the vase on condition of withdrawing his opposition. The consequence was that the Duchess became the purchaser at the price of 1800 guineas. Wedgwood then made fifty copies of the vase, which he sold at fifty guineas each; he is said to have paid £400 for the model and the entire cost of producing the copies is stated to have exceeded the sum received by him. Sir Joseph Banks and Sir Joshua Reynolds bore testimony to the excellent execution of these copies.

The fame of Wedgwood's operations was such, that his works at Burslem, and subsequently at Etruria, a village built by him near Newcastle-under-Lyne, and to which he removed in 1771, became a point of attraction to visitors from all parts of Europe.

The principal of the species of earthenware and porcelain invented by Wedgwood, according to Chalmers, are: 1, A terra cotta resembling porphyry, granite, Egyptian pebble, and other beautiful stones of the siliceous or crystalline order; 2, Basalts or black ware, a black porcelain biscuit of nearly the same properties with the natural stone, receiving a high polish, resisting all the acids, and bearing without injury a very strong fire; 3, White porcelain biscuit, of a smooth, wax-like appearance, of similar properties with the preceding; 4, Jasper, a white porcelain of exquisite beauty, possessing the general properties of basalts, together with the singular one of receiving through its whole substance, from the admixture of metallic calces, the same colours which those calces give to glass or enamels in fusion—a property possessed by no porcelain of ancient or modern composition; 5, Bamboo, or cane-coloured biscuit porcelain, of the same nature as the white porcelain biscuit; and 6, A porcelain biscuit remarkable for great hardness, little inferior to that of agate—a property which, together with its resistance to the strongest acids, and its impenetrability to every known liquid, renders it well adapted for the formation of mortars, and many different kinds of chemical vessels. The above six distinct species of ware, together with the "queen's ware" first noticed, have increased by the industry and ingenuity of different manufacturers, and particularly by Wedgwood himself and his son, into an almost endless variety of forms for ornaments and use.

Josiah Wedgwood closed a life of useful labour on 3rd January, 1795, in the sixty-fourth year of his age. Having acquired a large fortune, his purse was always open to the calls of charity, and to the support of every institution for the public good. To the poor he was a benefactor in the most enlarged sense of the word, and by the learned he was highly respected for his original genius and persevering industry in plans of the greatest national importance.

HENRY CORT.

In the manufacture of iron the first product of the blast-furnace is pig or cast iron—that is, iron in combination with carbon and silicon. To reduce this to a state of malleable iron, two methods are employed. One, the older, consists in the exposure of the melted pig-iron in a finery or hearth to the highly oxiding action of a blast of atmospheric air. The other, the modern practice, consists in stirring the melted pig-iron on the bed of a reverberatory furnace, so as to bring each portion of the whole mass successively up to the surface, and allow the oxygen of the air to seize upon and combine with the carbon and silicon, which become separated from the iron in the form of "cinder," leaving the product of the operation malleable or "wrought iron." This last process is termed "puddling," and the invention of it is usually ascribed to Henry Cort, as well as the method of producing bar-iron, by means of grooved rolls instead of by the old process of beating it out by forge hammers.

As in the case of most inventions, Cort's claim has been disputed; and Dr. Percy, in his work on *Metallurgy*, shows that other inventors are at least entitled to share in the merit, more particularly the Craneges of Coalbrookdale, and Peter Onions of Merthyr-Tydvil, both of whose patents preceded Cort's. But it does not appear that the inventions of either the Craneges or Onions were adopted by iron-makers to any large extent, and the merit certainly belongs to Henry Cort of practically introducing the method of puddling and manufacturing iron now generally followed, and which may be said to have established quite a new era in the history of the iron manufacture. When Cort took out his patent, the quantity of pig-iron produced in England was about 90,000 tons a-year; in 1866 it was above 4,000,000 tons. It was said in that year that there were no fewer than 8200 of Cort's furnaces in operation in Great Britain alone.

The story of Henry Cort is well and impartially told by Dr.

Percy. Cort was so unfortunate or so unwise as to become connected as partner with one Samuel Jellicoe, son of the Deputy-Paymaster of Seamen's Wages. To enable the firm to carry on their business, the elder Jellicoe advanced to them large sums out of the public moneys lodged in his hands; and when his accounts were investigated, it was found that the Cort partnership owed him, or rather the Treasury, upwards of £27,000. As Cort had assigned his patents to Jellicoe as security for the advances, they were at once taken possession of by the Crown; but although the processes which formed the subject of the patents were very shortly adopted to a large extent by the Welsh and other iron-masters, the Government never levied any royalty for their use, and the whole benefit of the inventions was thus made over to the public. Had Cort's estate been properly handled, there is every reason to believe that not only would the debts due by him to the Treasury have been paid, but that Cort himself would have realised a handsome fortune. As it was, the Government lost the money owing to the public treasury, while Cort was consigned to total ruin.

"This story," says Dr. Percy, "is one of the saddest in the annals of invention. Cort died in poverty, though he laid the foundation of the riches of many an iron-master, and largely contributed to the development of the resources and wealth of Great Britain. It is true that the value of the process of puddling has been greatly enhanced by subsequent improvements, especially two, viz., the application of iron bottoms to the puddling-furnaces, and the boiling process. But this has been the course with many inventions—perfection only being arrived at by slow degrees; and merit is not the less because others subsequently appear who improve the methods of their predecessors."

SAMUEL CROMPTON.

The history of Samuel Crompton and his famous invention of the mule spinning-machine is an extremely interesting one. Crompton was born on the 3rd of December, 1753, at Firwood, near Bolton. His father was a farmer, and the household, after the custom of Lancashire in those days, employed their leisure in carding, spinning, and weaving. Whilst Samuel was but a child, the family removed to a picturesque, old, rambling house, known as the Hole-in-the-Wood, about a mile from Bolton. Shortly afterwards, the father died.

Widow Crompton was a strong-minded woman, and carried on her husband's business with energy and thrift. She was noted for her excellent butter, honey, and elder-berry wine. When her son was about sixteen years old, she set him to earn his living by spinning at home, and exacted from him a certain amount of work daily. His youth at Hall-in-the-Wood was passed in comparative seclusion. All day he was alone at work, his mother doing the bargaining and fighting with the outer world.

It was with one of Hargreaves' jennies that Crompton span. The yarn was soft, and was constantly breaking; and if the full quantity of allotted work was not done, Mrs. Crompton scolded, and the time spent in mending broken threads kept him from his books and music, for he was a great reader, and a skilful player on the violin. Much annoyance of this kind drove his ingenuity into the contrivance of some improvements.

Five years—from his twenty-first year in 1774 to his twenty-sixth in 1779—were spent in the construction of the mule "My mind," he relates, " was in a continual endeavour to realise a more perfect principle of spinning; and though often baffled, I as often renewed the attempt, and at length succeeded to my utmost desire, at the expense of every shilling I had in the world." He was, of course, only able to work at the mule in the leisure left after each day's task of spinning, and often in hours stolen from sleep. The purchase of tools and materials absorbed all his spare cash, and when the Bolton Theatre was open, he was glad to earn eighteenpence a-night by playing the violin in the orchestra. The first mule was made, for the most part, of wood; and to a small roadside smithy he used to resort, "to file his bits o' things."

Crompton proceeded very silently with his invention. Even the family at Hall-in-the-Wood knew little of what he was about, until his lights and noise, while at work in the night time, excited their curiosity. Besides, inventors of machinery stood in great danger from popular indignation. The Blackburn spinners and weavers had driven Hargreaves from his home, and destroyed every jenny of more than twenty spindles for miles round.

When this storm was raging, Crompton took his mule to pieces, and hid the various parts in a loft or garret near the clock in the old Hall. Meanwhile, he excited much surprise in the market by the production of yarn, which, alike in fineness and firmness, surpassed any that had ever been seen. It immediately became the universal question in the trade, How does Crompton make that yarn? It was at once perceived that the greatly-desired muslins, brought all the way from the East Indies, might be woven at home, if only such yarn could be had in abundance.

At this time Crompton married, and commenced housekeeping in a cottage near the Hall, but still retained his workroom in the old place. His wife was a first-rate spinner, and her expertness, it is said, first drew his attention to her. Orders for his fine yarn, at his own prices, poured in upon him; and though he and his young wife spun their hardest, they were quite unable to meet a hundredth part of the demand. Hall-in-the-Wood became besieged with manufacturers, praying for supplies of the precious yarn, and burning with desire to penetrate the secret of its production. All kinds of stratagems were practised to obtain admission to the house. Some climbed up the windows of the workroom and peeped in; Crompton set up a screen to hide himself, and even that was not sufficient. One inquisitive adventurer is said to have hid himself for some days in a loft, and to have watched Crompton at work through a gimlet hole in the ceiling.

If Crompton had only possessed a mere trifle of worldly experience, there is no reason why, at this juncture, he might not have made his fortune. Unhappily, his seclusion and soft disposition placed him as a babe at the mercy of sharp and crafty traders. He discovered he could not keep his secret. "A man," he wrote, "has a very insecure tenure of a property which another can carry away with his eyes. A few months reduced me to the cruel necessity either of destroying my machine, or of giving it to the public. To destroy it, I could not think of; to give up that for which I had laboured so long, was cruel. I had no patent, nor the means of purchasing one. In preference to destroying, I gave it to the public."

Many, perhaps the majority of inventors, have lacked the means to purchase a patent, but have, after due inquiry, usually found some capitalist willing to provide the requisite funds. There seems no doubt that, had Crompton had the sense to bestir himself, he could easily have found a friend to assist him in securing a patent for the mule, or the Hole-i'-th'-Wood-wheel, as the people at first called it.

He says he "gave the mule to the public;" and virtually he did, but in such a way that he gained no credit for his generosity, and was put to inexpressible pain by the greed and meanness of those with whom he dealt. Persuaded to give up his secret, the following document was drawn up:—

"We, whose names are hereunto subscribed, have agreed to give, and do hereby promise to pay unto, Samuel Crompton, at the Hole-in-the-Wood, near Bolton, the several sums opposite our names, as a reward for his improvement in spinning. Several of the principal tradesmen in Manchester, Bolton, etc., having seen his new machine, approve of it, and are of opinion that it would be of the greatest public utility to make it generally known, to which end a contribution is desired from every well-wisher of trade."

To this were appended fifty-five subscribers of a guinea each, twenty-seven of half-a-guinea, one of seven shillings and sixpence, and one of five shillings and sixpence; making, together, the munificent sum of £72, 11s. 6d., or less than the cost of the model mule which Crompton gave up to the subscribers. Never, certainly, was so much got for so little. The merciless transaction receives its last touch of infamy from the fact recorded by Crompton in these words:—"Many subscribers would not pay the sums they had set opposite their names. When I applied for them, I got nothing but abusive language to drive me from them, which was easily done; for I never till then could think it possible that any man could pretend one thing and act the direct opposite. I then found it was possible, having had proof positive."

Deprived of his reward, Crompton devoted himself steadily

to business. He removed to Oldhams, a retired place two miles to the north of Bolton, where he farmed several acres, kept three or four cows, and span in the upper storey of his house. His yarn was still the best and finest in the market, and, as a consequence, he was plagued with visitors, who came prying about, under the idea that he had effected some improvement in his invention. His servants were continually bribed away from him, in the hope that they might be able to reveal something that was worth knowing. Sir Robert Peel (the first baronet) visited him at Oldhams, and offered him a situation with a large salary, and the prospect of a partnership; but Crompton had a morbid dislike to Peel, and he declined the overtures which might have led to his lasting comfort and prosperity.

Aided by the mule, the cotton manufacture prodigiously developed itself; but thirty years elapsed ere any serious attempt was made to recompense the ingenuity and perseverance to which the increase was owing. At last, in 1812, it was resolved to bring Crompton's claim before Parliament. It was proved that 4,600,000 spindles were at work on his mules, using up 40,000,000 lbs. of cotton annually; that 70,000 persons were engaged in spinning; and 150,000 more in weaving the yarn so spun; and that a population of full half-a-million derived their daily bread from the machinery his skill had devised. The case was clear, and Mr. Percival, the Chancellor of the Exchequer, was ready to propose a handsome vote of money, when Crompton's usual ill-luck intervened in a most shocking manner.

It was the afternoon of the 11th of May, 1812, and Crompton was standing in the lobby of the House of Commons, conversing with Sir Robert Peel and Mr. Blackburne, when one of them observed, "Here comes Mr. Percival." The group was instantly joined by the Chancellor of the Exchequer, who addressed them with the remark, " You will be glad to know that I mean to propose £20,000 for Crompton ; do you think it will be satisfactory?" Hearing this, Crompton moved off from motives of delicacy, and did not hear the reply. He was scarcely out of sight, when the madman Bellingham came up and shot Percival dead. This frightful catastrophe lost Crompton £15,000. Six weeks intervened before his case could be brought before Parliament, and then, on the 24th June, Lord Stanley moved that he should be awarded £5000, which the House voted without opposition; £20,000 might have been had as easily, and no reason appears to have been given for the reduction of Mr. Percival's proposal. All conversant with Crompton's merits felt the grant to be inadequate, whether measured by the intrinsic value of his service, or by the rate of rewards accorded by Parliament to other inventors.

HENRY BELL.

Steam navigation was introduced on American waters in 1807, Fulton launching his steamboat on the Hudson on the 3rd of October of that year. It was not, however, till 1812 that the first regular passenger steamer made its appearance in this country on the Clyde. This was the *Comet*, built for Mr. Henry Bell, the proprietor of the Helensburgh baths on the Clyde, and who had long been a most zealous advocate of steam propulsion.

Henry Bell was born in Linlithgowshire, in 1767. Dr. Cleland, in his work on Glasgow, speaks of him as an " ingenious, untutored engineer and citizen of Glasgow," and states that it may be said, without the hazard of impropriety, that Mr. Bell "invented" the steam-propelling system, "for he knew nothing of the principles which had been so successfully followed out by Mr. Fulton."

The construction of the *Comet* was begun in 1811, and the boat was so named in consequence of the appearance of a large comet that year. Mr. Bell was his own engineer, and in January, 1812, the first trial took place on the Clyde.

The little vessel was forty feet long on the keel, and ten feet six inches beam, propelled by a steam-engine of three or four horse-power, with a vertical cylinder, and working on the bell-crank principle—the engine being placed on one side of the vessel, and the boiler, of wrought-iron, on the other. She had two small paddle-wheels on each side, each wheel having four boards only.

For some time the *Comet* plied regularly between Glasgow and Greenock, at a speed of about five miles an hour. She was afterwards transferred to the Forth, where she ran for many years between the extremity of the Forth and Clyde Canal and Newhaven, near Edinburgh. The distance is 27 miles, and is stated by Mr. Bell to have been performed, on the average, in $3\frac{1}{2}$ hours, being at the rate of above $7\frac{1}{2}$ miles an hour.

Mr. Bell's experiments did not realise to himself those pecuniary advantages which were due to his enterprise. From the city of Glasgow he received in his latter years a small annuity, in acknowledgment of his services to commerce and civilisation. He died at Helensburgh, on the Clyde, in 1830. A monument was erected to his memory near Bowling.

SIR DAVID BREWSTER.

In the whole history of science there is not, perhaps, any discovery of ancient or of modern date that promised so rich a reward to the inventor, and was so completely anticipated, as in the case of the kaleidoscope. The very announcement of the patent, by which the discovery was intended to be secured, was immediately followed by an infringement so extensive as to leave all legal redress unattainable. But the piracy did not terminate here; for various attempts were made to deprive its author, Sir David Brewster, of the merits of the discovery, and to refer it to Baptista Porta, Harris, Wood, Bradley, &c. All these were very satisfactorily answered by Sir David Brewster, confirmed by Professor Playfair, Mr. Watt, and Professor Pictet, who attested the originality of the invention.

It was in the year 1814, when Sir David Brewster was engaged in experiments on the polarization of light by successive reflections between plates of glass, the reflectors being in some cases inclined to each other, that he had occasion to remark the circular arrangement of the images of a candle round a centre, or the multiplication of the sectors formed by the extremities of the glass plates. In repeating at a subsequent period the experiments of M. Biot on the action of fluids upon light, Brewster placed the fluids in a trough formed by two plates of glass cemented together at an angle. The eye being necessarily placed at one end, some of the cement which had been pressed through between the plates appeared to be arranged into a regular figure. The symmetry of this figure being very remarkable, Sir David Brewster set himself to investigate the cause of the phenomenon, and in doing this he discovered the leading principles of the kaleidoscope.

Upon these principles Sir David Brewster constructed an instrument, in which he fixed *permanently*, across the ends of the reflectors, pieces of coloured glass and other irregular objects; but the great step towards the completion of the instrument remained yet to be made, and it was not till some time afterwards that the idea occurred to the inventor of *giving motion to objects*, such as pieces of coloured glass, &c., which were placed loosely in a cell at the end of the instrument. When this idea was carried into execution, the kaleidoscope in its simple form was completed.

The next, and by far the most important step of the invention, was to employ a draw-tube and lens, by means of which beautiful forms could be created from objects of all sizes, and at all distances from the observer. In this way the power of the kaleidoscope was indefinitely extended, and every object in nature could be introduced into the picture, in the same manner as if these objects had been reduced in size, and actually placed at the end of the reflectors.

The kaleidoscope being now completed, Brewster was urged by his friends to secure the exclusive property of it. After the patent was signed, and the instruments in a state of forwardness, the gentleman who was employed to manufacture them under the patent, carried one to show to the principal London opticians, for the purpose of taking orders for them. These gentlemen naturally made one for their own use and the amusement of their own friends; and the character of the instruments being thus made public, the manufacture extended to tinmen and glaziers; and kaleidoscopes were soon hawked about the streets of London at all prices, some even as low as a shilling. No proof of the originality of the kaleidoscope could be stronger than the sensation which it created in London and Paris. In the memory of man, no invention and no work, whether addressed to the imagination or the understanding, ever produced such an effect. A universal mania for the instrument seized all classes, from the lowest to the highest, from the most ignorant to the most learned; and every person not only felt, but expressed the feeling, that a new pleasure had been added to their existence.

The pirated instruments, of course, were only of the simple form, and necessarily of rude and unscientific construction. They, however, had the effect of deeply injuring the property of the inventor; but the rage was soon over, and they were thrown aside as a pleasing but useless toy.

This, however, is not the case with the patent kaleidoscope, which is of great service in exhibiting an infinite variety of beautiful patterns, which are transferred to several of our manufactures. The system of endless changes is named as one of the most astonishing properties of the kaleidoscope. With a number of loose objects,—pieces of glass, for example—it is impossible to reproduce any figure we have admired, when it is once lost; centuries may elapse before the same combination returns. If the objects, however, are placed in the cell so as to have very little motion, the same figure may be recalled; and, if absolutely fixed, the same pattern will return in every evolution of the object plate. A calculation of the number of forms is given upon the ordinary principles of combination—namely, that 24 pieces of glass may be combined

$$1391724288887255299942512849340 2200$$

times, an operation the performance of which would take hundreds of thousands of millions of years, even upon the supposition that twenty of them were performed every minute. This calculation, surprising as it appears, is false, not from being exaggerated, but from being far inferior to the reality. It proceeds upon the supposition that one piece of glass can exhibit only one figure, and that two pieces can exhibit only two figures; whereas it is obvious that two pieces, though they can only be combined in two ways on the same straight line, yet the one can be put above and below the other, as well as on its right side or its left side, and may be joined so that the line connecting those centres may have an infinite number of positions with respect to an horizontal line.

CHARLES BABBAGE.

The calculating machines of the late Mr. Babbage have at different times excited much interest on the part of the public and of scientific men.

" Mr. Babbage," says a writer in the *Encyclopædia Britannica*, "was a fellow student at Cambridge with Sir John Herschel and Dean Peacock, and along with them contributed by his writings and personal efforts to introduce into that university the improved Continental mathematics.

" A few years after leaving college, he originated the plan of a machine for calculating tables by means of successive orders of differences; and having received for it in 1822 and the following year the support of the Astronomical and Royal Societies, and a grant of money from Government, he proceeded to its execution. It is believed that Mr. Babbage was the first who thought of employing mechanism for computing tables by means of differences: the machine was subsequently called the *difference machine.*'

" In the course of his proceedings, Mr. Babbage invented a mechanical notation (described in the *Philosophical Transactions* for 1826), intended to show the exact mutual relations of all the parts of any connected machine, however complex, at any given instant of time. He also made himself acquainted with the various machines used in the arts, with the tools used in constructing them, and with the details of the most improved workshops. Employing Mr. Clements, a skilful mechanist, a portion of the calculating machine, very beautifully constructed, was brought into working order, and its success so far answered the expectations of its projector. But, notwithstanding several additional grants from Government, the outlay on this most expensive kind of work soon exceeded them. The part actually constructed is now placed in the museum of King's College, London. It employs numbers of nineteen digits, and effects summations by means of three orders of differences. Though only constituting a small part of the intended engine, it involves the principles of the whole. The inventor proposed to connect it with a printing apparatus, so that the engine should not only tabulate the numbers, but also print them beyond almost the possibility of error.

"At this stage (1834), Mr. Babbage contrived a machine of a far more comprehensive character, which he called the Analytical Machine, extending the plan so as to develop algebraic quantities, and to tabulate the numerical value of complicated functions when one or more of the variables which they contain are made to alter their values. Had this machine been constructed, it would necessarily have superseded what had already been done. Government were not unnaturally startled by this new proposal; and as about the same time Mr. Babbage's relations to Mr. Clements were broken off, the difficulties of the affair became insurmountable.

"'The opinions of men of science are not unanimous as to the great practical importance of calculating tables by machinery; but the improvements of mechanical contrivance, which the joint skill of Mr. Babbage and Mr. Clements introduced into engineering workshops, are unquestionably of great importance to the arts."

Mr. Babbage was born in 1790, and died on the 18th of October, 1871. He was the author of several valuable works. One, "On the Economy of Manufactures and Machinery," published in 1832, has gone through several editions, and been translated into several languages. In it all mechanical processes are classified from the most scientific point of view, and the most interesting examples of the more important kinds of manufactures are described. In addition to this work, we may mention his "Comparative View of the different Life Assurance Societies," his "Differential and Integral Calculus," his "Decline of Science, A Ninth Bridgewater Treatise," and "The Exposition of 1851."

HENRY BESSEMER

The invention of the Bessemer process of decarburising pig-iron while in a molten state, by blowing atmospheric air through it, and thereby producing steel, is an interesting story. Mr. Bessemer's discovery was in some measure accidental, like so many other discoveries in the arts. The remarkable thing is, that, taking into consideration the attention paid to the chemistry of metallurgy of late years, the discovery was not made long ago; and that it should have been reserved for Mr. Bessemer to make it, who was neither a chemist nor an iron manufacturer.

About 1856, says a writer in the *Quarterly Review*, the minds of inventors were running in the direction of improved guns. It was believed that these might be made much stronger if some better material than cast-iron were used; and Mr. Bessemer, like many others, began a series of experiments to solve the problem if he could. He first tried a mixture of cast-iron and cast-steel, the result being a half-decarbonised cast-iron. Guns made of this metal were found to possess great strength; but as they were of comparatively small bore, 24-pounders, Mr. Bessemer resolved to make them on a larger scale, for the purpose of more exclusively testing the strength of the material.

In the course of his experiments, the idea occurred to him that if he could contrive to blow air through melted pig-iron, he would be enabled to purify it to an unusual extent. He thought that, by thus bringing oxygen into contact with the fluid metal, the carbon with which it was surcharged would be removed, as well as the silicon, phosphorous, and sulphur which it contained. This is exactly what is done, after another and very laborious method, in the process of puddling. He proposed to reverse this process, and so get rid of puddling altogether. Instead of bringing the particles of the iron in turn into contact with the oxygen of the air, his scheme was to force the air through the fluid mass into contact with the separated particles of the iron. Now that the thing is done, we see how simple, how natural the first idea was. But it needs the quick intuition of genius to detect even simple things in practical science.

The only way of determining the matter was by putting the idea to the test of experiment. Accordingly, early in 1856, Mr. Bessemer ordered a stock of Blaenavon iron, and set up a blast-engine and cupola at Baxter House, St. Pancras, where he then resided.

The first apparatus which he used for conversion was a fixed cylindrical vessel three feet in diameter and four feet high, somewhat like an ordinary cupola furnace, lined with fire-bricks; and at some two inches from the bottom he inserted five twyer pipes, with orifices about three-eighths of an inch in diameter. About half-way up was a hole for running in the molten metal, and on the opposite side at the bottom was the tap-hole, by which the metal was to be run off at the end of the process.

The first experiment was not made without occasioning considerable alarm. It was a most unusual process, and it looked dangerous, as indeed it proved to be. When the charge of pig-iron was melted, the blast was turned on to prevent its running into twyer holes, and then the fluid metal was poured in through the charging-hole by the attending stoker. A tremendous commotion immediately took place within the vessel. the molten iron bounded from side to side; a violent ebullition was heard going on within; while a vehement violet-coloured flame, accompanied with dazzling sparks, burst from the throat of the cupola, from which the slag was also ejected in large foam-like masses. A cast-iron plate, of the kind used to cover holes in the pavement, that had been suspended over the mouth of the vessel, dissolved in a gleaming mist, together with half-a-dozen yards of the chain by which it hung. The air-cock was so close to the vessel that no one durst go near to turn it and stop the process. The flames shot higher and higher, threatening the destruction of the building, and the fire-engines were sent for in hot haste. Before they arrived, however, the fury of decarbonisation had expended itself, and the product was run off.

The result was not quite satisfactory: the product was for the most part "burnt" iron; but the experiment was sufficiently encouraging to induce Mr. Bessemer to make a second trial, and the product was found to be malleable iron.

In the course of further experiments, it was found that, by interrupting the process before the decarburisation of the iron was complete, the product was unmistakable steel, which was tried and found of good quality.

Here was a discovery of immense importance. If malleable iron and steel could be thus made direct from pig-iron by a process so rapid and simple, it could not fail before long to effect an entire revolution in the iron trade.

The news of Mr. Bessemer's discovery soon flew abroad, and many distinguished metallurgists went to see the process. Among others, Dr. Percy went, and thus describes what he saw:—

"Towards the end of 1856, I had the pleasure of seeing the process in operation at Baxter House, and I confess I never witnessed any metallurgic process more startling and impressive. After the blast was turned on, all proceeded quietly for a time, when a volcano-like eruption of flame and sparks suddenly occurred, and bright red-hot scoriæ or cinders were forcibly ejected, which would have inflicted serious injury on any unhappy bystanders whom they might perchance have struck. After a few minutes, all was again tranquil, and the molten, malleable iron was tapped off."

Though the doctor came away wondering, he was not convinced. He analysed a portion of the iron which he had seen produced; and when he found it to contain 1 per cent, of phosphorus, he says his scepticism was rather confirmed than otherwise.

Among other visitors at Baxter House was George Rennie, the engineer, who, after witnessing the process, urged Mr. Bessemer to draw up an account of it for the meeting of the British Association at Cheltenham in the autumn of 1856. To this the inventor assented, and the result was his paper "On the Manufacture of Iron and Steel without Fuel."

On the morning of the day on which the paper was to be read, Mr. Bessemer was sitting at breakfast in his hotel, when an ironmaster (to whom he was unknown) said, laughing, to a friend within his hearing, "Do you know there is some one come down from London to read a paper on making steel from cast-iron without fuel! Did you ever hear such rubbish?" The ironmaster, however, was of a different opinion as to the new invention after he had heard the paper read. Its title was certainly a misnomer, but the correctness of the principles on which the pig-iron was converted into malleable iron, as explained by the inventor, was generally recognised, and there seemed to be good grounds for anticipating that the process would before long come into general use.

The *rationale* of the method of conversion was intelligible and simple. Mr. Bessemer held that, by forcing atmospheric air through the fluid metal, the oxygen was brought into contact with the several particles of the iron and carbon, combining with the latter to form carbonic-acid gas, which passed off by the throat of the vessel, through which the slag was also ejected, leaving as the product, when the combustion was complete, a mass of malleable iron, which was run off by the tap, into the ingot moulds placed to receive it. "Thus," said he, "by a single process, requiring no manipulation or particular skill, and with only one workman from three to five tons of crude iron pass into the condition of several piles of malleable iron in from thirty to thirty-five minutes, with the expenditure of about one-third part of the blast now used in a fiery furnace with an equal charge of iron, and with the consumption of no other fuel than is contained in the crude iron."

In the same paper the inventor called attention to an important feature of the new process in the following words :—" At the stage of the process immediately following the boil, the whole of the crude iron has passed into the condition of cast-steel of ordinary quality. By the continuation of the process, the steel so produced gradually loses its small remaining portion of carbon, and passes successively from hard to soft steel, from softened steel to steely iron, and eventually to very soft iron; hence, at any certain point of the process, any quality may be obtained."

It was, however, found in practice that this remarkable peculiarity of the Bessemer process constituted its principal defect. Thus it was extremely difficult, if not impracticable, to determine with certainty when the decarburisation had proceeded to the desired extent, and to the exact point at which the blast was to be stopped. If arrested too soon, no dependence could be placed on the result, as the metal might be only one-half or three-fourths converted, according to chance; while, if continued till the iron was quite decarburised, it would be burnt and comparatively worthless. The workmen could only judge by the appearance of the flame—first violet, then orange, then white—issuing from the mouth of the vessel, when it was proper to interrupt the process. But the eyesight of the workmen was not to be depended on; and as the stoppage of the blast ten seconds before or ten seconds after the proper point had been attained would produce an altogether different result, it began to be feared that, on this account, the Bessemer process, however ingenious, could never come into general use. Indeed, the early samples of Bessemer steel were found to exhibit considerable irregularity: the first steel tyres made of the metal, tried on some railways, were found unsafe, and their use was abandoned; and the ironmasters generally, who were of course wedded to the established processes, declared the much-vaunted Bessemer process to be a total failure. It was regarded as a sort of meteor that had suddenly flitted across the scientific horizon, and gone out leaving the subject in more palpable darkness than before.

Mr. Bessemer himself was by no means satisfied with the result of his first experiments. He was satisfied that he had hit upon the right principle; the question was, could he correct those serious defects in the process which to practical men seemed to present an insuperable obstacle in the way of the adoption of his invention?

It was a case for persevering experiment, and experiment only. The inventor's patience and perseverance were found equal to the task. Assisted by Mr. Longsdon, he devoted himself for several years to the perfection of his process of conversion, in which he at last succeeded.

We can only briefly refer to the alterations and improvements in the mode of conducting it which he introduced. In the first place, he substituted for the fixed converting vessel, a movable vessel, mounted on trunnions, supported on stout pedestals, so that a semi-rotatory motion might be communicated to it at pleasure. It was found both dangerous and difficult, while the converting vessel was fixed, to tap the cupola furnace; for the blast had to be continued during the whole time the charge was running out of the vessel, in order to prevent the remaining portion from entering the twyers. By the adoption of the movable converting vessel, this source of difficulty was completely got rid of, while the charging of the vessel with the fluid metal, the interruption of the process at the precise moment, and the discharge of the metal when converted, were rendered comparatively easy. The position and action of the twyers were also improved, and the converting vessel was lined with "ganister," a silicious stone, capable of resisting the action of heat and slags, so as to last for nearly a hundred consecutive charges before becoming worn out, whereas the lining of firebrick, originally used, was ordinarily burnt out in two charges of twenty minutes each.

Another important modification in the process related to the kind of metal subjected to conversion, and its after treatment. In his earliest experiments, Mr. Bessemer had, by accident, made use of a pure Blaenavon iron; but in his subsequent trials, iron of an inferior quality had been subjected to conversion, and the results were much less satisfactory. It was found that the high temperature and copious supply of air blown through the metal had failed to remove the sulphur and phosphorus present in the original pig, and that the product was an inferior metal. After a long series of experiments, Mr. Bessemer at length found that the best results were obtained from Swedish, Whitehaven, Hæmatite, Nova Scotian, or any other comparatively pure iron, which was first melted in a reverberatory furnace, before subjecting it to conversion, in order to avoid contamination by the sulphur of the coal.

Finally, to avoid the risk of spoiling the metal while under conversion by the workmen stopping the blast at the wrong time, Mr. Bessemer adopted the method of refining the whole contents of the vessel by burning off the carbon, and then introducing a quantity of fluid carburet of iron, containing the exact measure of carbon required for the iron or steel to be produced.

When Mr. Bessemer, after great labour and expense, had brought his experiments to a satisfactory issue, and ascertained that he could produce steel of a quantity and texture that could be relied on with as much certainty as any other kind of metal, he again brought the subject of his invention under the notice of the trade; but, strange to say, not the slightest interest was now manifested in it. The Bessemer process had been set down as a failure, and the iron and steel makers declined to have anything to do with it. The inventor accordingly found that either the invention must be abandoned, or he himself must become steel manufacturer. He adopted the latter alternative, and started his works in the very stronghold of steelmaking, at Sheffield, with a success which is matter of history.

The great value of this invention is unmistakably shown by the fact that 500,000 tons of steel were, in 1874, being made annually by the Bessemer process in Great Britain, the total number of converting vessels in use being ninety-one, and their aggregate capacity 467 tons. Large quantities are also manufactured by it in Sweden, Russia, Austria, Prussia, and other European countries. In America it is likewise extensively employed.

A recent experimental trial is reported, which is said to have been quite fairly conducted: as the result, it was found that a Bessemer steel rail lasted fully longer than twenty iron ones.

JOHN ERICSSON.

IN a mountain hamlet in a beautiful district of Central Sweden, near the iron mines of Langbanshyttan, stands in front of a miner's cottage a granite-shaft eighteen feet high, with this inscription in golden letters—

JOHN ERICSSON

WAS BORN HERE IN 1803.

September 3, 1867, when this monument was unveiled, was a gala day in the mining districts; all work was suspended, and crowds gathered from far and near to do honour to their distinguished countryman, whose fame is world-wide, and of whom they may well be proud.

John Ericsson is the son of Olof, a Swedish miner; his mother was a woman of intelligence and refinement; John's brother, Nils, distinguished himself highly in his native country as an engineer of canals and railroads, and was raised to the rank of Baron by the King of Sweden.

John Ericsson's genius developed early. His childish toys were machinery and tools of his own construction. At nine years of age he saw through one of the draught offices on the grand ship canal of Sweden, and there caught some idea of how to use the drawing instruments. This knowledge was put to use in most ingenious ways. Accompanying his father to a pine forest, where he went to select timber for lock gates on the canal, the miniature draughtsman set himself to construct a pair of compasses of birch wood, with needles stuck in the ends of the legs; his good-natured mother allowing him to convert a pair of steel tweezers, taken from her dressing-case, into a drawing pen, also to take hair from her sable-cloak for paint brushes.

It may be mentioned that Ericsson's grandfather by the mother's side was a man of property; which, however, was lost in some mining speculations, so that young Ericsson began life in poverty, which was perhaps fortunate for the development of his genius.

Before he was eleven Ericsson had constructed a sawmill and planned a pumping-engine for the purpose of clearing the mines of water. He had much difficulty in perfecting this plan, as it was intended to be operated by a windmill, which he had never seen; at last his father visited one, and after hearing his description he was able to finish it.

This may be said to be the turning-point in his career. The plan was shown to Admiral Count Platen, President of the Gotha Ship Canal, on which Ericsson's father was then employed. The Count, amazed at the genius displayed by such a mere child, uttered the encouraging and prophetic words, "Go on as you have begun, and you will one day produce something extraordinary." He was appointed a cadet in the Swedish corps of mechanical engineers at the age of twelve, and at thirteen was appointed a *nevelleur* (leveller) on the part of the canal over which Count Platen presided. Six hundred troops working on this canal were directed by this boy, and it is amusing to note that one of his attendants carried a stool to enable the small engineer to reach his levelling instruments.

Many important works on the canal were constructed from

Ericsson's plans at this early period. His contact with military men incited him, much against Count Platen's will, to become a soldier, and he entered the army as an ensign. He was very soon promoted to a lieutenancy.

The posts of government surveyors being open to competition by military officers, Ericsson entered his name as a competitor, and of course he gained a prize. His industry in his new employment was most untiring; maps which he executed at this period are still preserved at Stockholm. Not satisfied with doing more work than others in this employ, he directed his energies to the drawing and engraving plates for the illustration of a work on the Gotha Canal. He invented a machine engraver for this work. He next turned his attention to the construction of a flame-engine. One was made with ten-horse power, and was successful. Ericsson now obtained leave to go to England; he arrived there in 1826 with his invention, and did not return to his native country. He sent in his resignation from the army, which was not accepted until he had been promoted to the post of captain, and then only with much regret. Alas! for the flame-engine; it proved a failure with the English sea-coal. He had to begin and experiment anew, and was at last successful in getting a machine constructed and patented and sold to John Braithwaite, who was a friend in need to the inventor. His inventions at this time are said to have been, "A pumping-engine on a new principle, engines with surface condensers and no smoke-stack, and blowers supplying the draught, applied to the steamship *Victory* in 1828. Apparatus for making salt from brine; mechanism for propelling boats on canals; a hydrostatic weighing machine, to which the Society of Arts awarded a prize; an instrument now in extensive use for taking soundings independently of the length of the lead-line; a file cutting machine, and various others, to the number of fourteen patented inventions and forty machines, all novel in design."

Ericsson entered, in 1829, into a competition for a prize of £500 offered by the Liverpool and Manchester Railway Company for the best locomotive. George Stephenson's engine, the *Rocket*, gained the prize. Ericsson's, the *Novelty*, was the swiftest, running at the speed of thirty miles an-hour, the speed of Stephenson's being only thirteen and a-half miles. Great enthusiasm was manifested over Ericsson's *Novelty*, but the judges decided in favour of the superior traction power of Stephenson's. The *Novelty* was planned and executed in the short space of seven weeks.

He next introduced the use of steam in fire-engines; one which was built for the King of Prussia, was the means of saving some valuable buildings in Berlin. In January, 1840, he won the gold medal of the New York Mechanics Institute for the best steam fire-engine.

One of his inventions, begun in England and carried on and completed in America, was the construction of the Caloric engine. This engine was fitted into a ship built for the purpose, and named the Ericsson, and of which the engineer himself says, "The ship after completion made a successful trip from New York to Washington and back during the winter season; but the average speed at sea proving insufficient for commercial purposes, the owners, with regret, acceded to my proposition to remove the costly machinery, although it had proved perfect as a mechanical combination.

"The resources of modern engineering having been exhausted in producing the motors of the Caloric ship, the important question has for ever been set at rest, Can heated air, as a mechanical motor, compete with steam.

"'The commercial world is indebted to American enterprise for having settled a question of such vital importance. The marine engineer has thus been encouraged to renew his efforts to perfect the steam-engine without fear of rivalry from a motor depending on the dilation of atmospheric air by heat."

Ericsson did not, however, give up the idea of the usefulness of this engine, as in places where water is scarce or not obtainable, it is necessary. It is gratifying to know that his exertions in this direction did not go unrewarded; the American Academy of Arts and Sciences voted in 1862: "That the Rumford premium be awarded to John Ericsson for his improvements in the management of heat, particularly as shown in his Caloric Engine."

Ericsson's tireless brain was now turned towards another possible motive-power, in the sun's rays. His solar engine was intended for use in sunny countries; as, for instance, Upper Egypt, North African deserts, Eastern Arabia, the western part of China, in the Western Hemisphere, Lower California, Mexico, and North America, &c. He can even now supply an engine of one hundred horse-power, to any one who can pay the price, but it is too expensive, under existing circumstances, to make the venture a profitable one. The time may come when it may be used with advantage. It is, indeed, a noble idea to convert a power which, in many countries, is a means of blight into a blessing. The screw propeller was suggested to Ericsson by the study of the movements of birds and fishes. This propeller was first offered to the English Admiralty, who rejected it, and thus England lost Ericsson; he met an American naval officer, who took up the idea with enthusiasm, and ordered two vessels to be made at his own expense. He removed to America in 1839. Captain Stocktin's government did not catch his enthusiasm so readily; and it was not till after many experiments and many delays, that must have been very irksome to the active Ericsson, that they finally adopted it

Ericsson completely revolutionised naval warfare; and, as in all revolutions of a like kind, many prejudices had to be overcome, in this case matters were hastened by the practical issue of a fight between Ericsson's boat the *Monitor* and the *Merrimack* in Hampton Roads, 9th March, 1862, resulting in the defeat of the latter. After this time the Monitors were adopted by Sweden, Norway, and Russia. England was the last to be convinced of their superiority. The introduction of vessels for conducting submarine warfare is now engaging the attention of the great engineer, and he has completed, at his own expense, the *Destroyer*, a vessel for discharging torpedoes. Ericsson deserved much at the hands of his adopted country; he himself asked little, and, we regret to say, got what he asked.

Honour has, however, been done to him by many distinguished scientific societies, both in Europe and America. In addition to the monument at his birthplace in Sweden, already mentioned, another was erected on the roadside near the ironworks of Langbanshyttan bearing his own and his brother's name and the words: "Both of whom have served and honoured their native land. Their way through work to knowledge and lasting fame is open to every Swedish youth." On the reverse side is the suggestive inscription: "The way to the schoolhouse of Langbanshyttan."

Although absent from his native land, it has not been forgotten by him. He built at his own cost, and presented to the Swedish Government, the machinery of a gun-boat as a model for a fleet of gun-boats, to be manœuvred by hand independently of steam. Of his personal history there is little to tell; his work represents his life; he does not go much into society, nor does he invite society to visit him at his home and workshop in Beach Street, New York. His amusements are found in his work. His habits are simple in the extreme, and unvaried; he takes gymnastic exercise in the morning after a cold bath, and walking exercise in the evening from ten till twelve. He is strictly temperate in his eating and drinking, and does not use tobacco; it says much for this mode of life when it is remembered that he is seventy-six, and is still vigorous in body and mind. Captain Ericsson is a widower and childless, but he has nephews living who have distinguished themselves in various ways.

It seems hard to believe that Ericsson was one of the pioneers in the early days of locomotives, that he competed against Stephenson, and was unsuccessful, and that in these later days he still leads the van of progress in the line of inventions which he has taken up.

THOMAS ALVA EDISON.

THOMAS ALVA EDISON was born at Milan in Erie county, Ohio, 11th February, 1847. It was, however, in Port Huron, Michigan, a village perhaps equally obscure, that the early youth of this wonderful genius was passed. His father is of Dutch descent; his mother was of Scotch parentage and born in Massachusetts, was well educated, having been a teacher in Canada, and from her Edison received instruction. Two months was all the time passed at a regular school. His father's occupation, or rather occupations, seem to have been both many and varied; by turns tailor, nurseryman, dealer in grain, in timber, and in farm lands. Possibly the versatility displayed by his distinguished son may be an inheritance from this parent. No precocious sayings or doings of young Edison are reported, but he was a great reader, devouring indiscriminately everything in the shape of a book that came in his way.

His first entrance into the world of work was made at the early age of twelve. He started as train boy on the Grand Trunk Railroad of Canada and Central Michigan. Here the peculiar bent of his mind began to show itself; in a disused corner of an old baggage car, where he kept his papers and other wares, he gradually accumulated a quantity of bottles and retort-stands, and, with the aid of "Fresenius's Qualitative

Analysis," tried some experiments in chemistry. An ingenious youth, he seems to have been peculiarly wide-awake in every direction.

His next original venture was the starting of a newspaper, printed in the train, and subscribed for by baggage-men and brakesmen. His mode of printing was certainly simple enough; he pressed the sheets on the type with his hands; this type he had purchased from the *Detroit Free Press* office, and there also, in some fashion or other, he had at spare moments picked up what he knew of type-setting. The *Grand Trunk Herald*, as his newspaper was named, would no doubt be in demand on his train. During the war he telegraphed his headlines to the country stations in advance. In addition to his acquisitions in practical matters, his reading tastes were diligently cultivated; indeed, he seriously set himself to the task of reading through the public library of Detroit. Happily for his future usefulness, this feat was an impossibility, although he made considerable progress in a shelf of dry and probably dusty volumes. It was not all progress with our lively youth, however. Through some accident, in his absence, among his bottles, the baggage waggon took fire, and the chemist's laboratory was ignominiously turned out by an enraged conductor, who finished up by thrashing the chemist. At another time he was made award of the responsibilities of editorship, by being thrown into the river by an aggrieved individual, who had been too personally attacked in another paper with which he was connected, named *Paul Pry*. We imagine he must have cost his parents some perturbation about this time, as he developed a mania for telegraphy, encircling his father's house with wires. His mode of obtaining material for his work was somewhat questionable—small boys, for a slight consideration would bring him quantities of zinc, taken from their respective homes. The station-master at Port Clements taught him telegraphing, in gratitude for the rescue of his child, Edison having snatched the infant from the front of an advancing train.

From this time he began to concentrate his energies and turned his attention more closely to electricity. He became thoroughly conversant with telegraphing, but as an ordinary operator did not succeed ; his duties were forgotten or neglected while he was deep in some experiment.

The strong inventive bent of his mind carried him away, and the close attention to the ordinary routine of work that was necessary seemed impossible to him, hence his wanderings from place to place.

Some of his small inventions succeeded about this time ; but his attempt to introduce a duplex system of telegraphy failed, and he went to New York with his ardour damped. The tide was about to turn in his favour, however. The Gold and Stock Company were fortunate enough to secure his services, and a useful invention of his which they adopted secured their favour.

The Western Union Telegraph Company also gave him a salary that they might get the benefit of his inventions in telegraphy. His success was now secured, and his inventive faculties had full play. At Newark he took regular lessons in chemistry. He married a Newark lady, Miss Mary Stillwell. In 1876 he removed to Menlo Park, an hour's ride from New York on the Pennsylvania Road. At a little distance from the village stands a long white wooden building of two stories. This contains the laboratory of this pioneer of a new age. Although it is twenty-eight by one hundred feet in extent, it is already too small for the needs of its owner. Many workmen and much machinery are employed in carrying out and perfecting the ideas of the inventor. Edison's personal appearance is thus described by a visitor who saw him at work: "A figure of perhaps five feet nine in height, bending intently above some detail of work. There is a general appearance of youth about it, but the face, knit into anxious wrinkles, seems old. The dark hair, beginning to be touched with gray, falls over the forehead in a mop. The hands are stained with acid, and the clothing is of an ordinary 'ready-made' order. He has the air of a mechanic, or more definitely, with his peculiar pallor, of a night-student. His features are large; the brow well shaped, without unusual development; the eyes light-gray; the nose irregular, and the mouth displaying teeth which are also not altogether regular. When he looks up his attention comes back slowly, as if it had been a long way off. But it comes back fully and cordially, and the expression of the face, now that it can be seen, is frank and prepossessing. A cheerful smile chases away the grave and somewhat weary look that belongs to it in its moments of rest. He seems no longer old. He has almost the air of a big careless schoolboy released from his tasks."

Such is Edison in work-hours, and it seems to be always work-hours with him; he carries on his researches far into the morning when he can have quiet, and it is then his most happy hits are made. With a sharp eye to everything which may be of service to him, no smallest discovery in chemistry is despised, but laid aside for probable use at some future time. Besides the phonograph and the practical adaptation of the electric light for purposes of illumination, the number and variety of his inventions are bewildering. Among his still uncompleted ideas may be mentioned the Megaphone, composed of a speaking-trumpet and two ear-trumpets, with which two people can converse in an ordinary tone some miles apart. Another is the æorophone, two hundred and fifty times the capacity of the human lungs, meant for shouting distances at sea. The sound is carried on in this instrument by a steam jet, which takes the tones of the voice and sends them on. Electric pens, electric shears, electric engines are a few among the many inventions of this wonderful genius. Edison seems to believe that the truest inspiration comes through work; he still goes on inventing, and he may yet astonish the world with an invention beyond our wildest dreams.

Edison's name is honourably connected with the desire to use the telephone as a means of sound communication, and the electric light a means of practical illumination. The telephone is being put to the test as a " far speaker," having been introduced as a means of communication in several business houses in London and elsewhere, while the electric light only requires some modification in the production and manipulation to be practically useful, at least, for the purposes of public illumination.

www.ingramcontent.com/pod-product-compliance
Lightning Source LLC
Chambersburg PA
CBHW022137300426
44115CB00006B/228